[英国]彼得·阿特金斯 著 傅兴阳 译

牛津通识读本·

化学

Chemistry

A Very Short Introduction

译林出版社

图书在版编目（CIP）数据

化学 / （英）彼得·阿特金斯（Peter Atkins）著；傅兴阳译. -- 南京：译林出版社, 2024. 10. -- （牛津通识读本）. -- ISBN 978-7-5753-0312-5

I. O6

中国国家版本馆CIP数据核字第2024W4Q173号

Chemistry: A Very Short Introduction by Peter Atkins
Copyright © Peter Atkins Limited 2013, 2015
Chemistry: A Very Short Introduction, First Edition was originally published in English in 2013. This licensed edition is published by arrangement with Oxford University Press. Yilin Press, Ltd is solely responsible for this bilingual edition from the original work and Oxford University Press shall have no liability for any errors, omissions or inaccuracies or ambiguities in such bilingual edition or for any losses caused by reliance thereon.
Chinese and English edition copyright © 2024 by Yilin Press, Ltd
All rights reserved.

著作权合同登记号　图字：10-2017-080 号

化学	[英国] 彼得·阿特金斯 / 著　傅兴阳 / 译
责任编辑	许　昆
装帧设计	景秋萍
校　　对	施雨嘉
责任印制	董　虎

原文出版	Oxford University Press, 2015
出版发行	译林出版社
地　　址	南京市湖南路 1 号 A 楼
邮　　箱	yilin@yilin.com
网　　址	www.yilin.com
市场热线	025-86633278
排　　版	南京展望文化发展有限公司
印　　刷	江苏凤凰通达印刷有限公司
开　　本	890 毫米 × 1260 毫米 1/32
印　　张	7
插　　页	4
版　　次	2024 年 10 月第 1 版
印　　次	2024 年 10 月第 1 次印刷
书　　号	ISBN 978-7-5753-0312-5
定　　价	39.00 元

版权所有·侵权必究

译林版图书若有印装错误可向出版社调换。质量热线：025-83658316

序 言

游书力

不知道从什么时候开始，自己经常会被问到是从何时开始接触化学，喜欢化学的。我特别希望我的答案是受到某一本书的启发，然后开启了化学之旅。很遗憾，这并没有在我的身上发生。但对于还处在科学启蒙阶段的学生来说，今天这本书应该很有机会提供这种缘分。而且对于任何一位想了解化学整体情况的读者，我都非常推荐这本言简意赅、深入浅出，又很好地兼顾了科普易懂与专业深度的通识书。

《化学》这本书的作者为彼得·阿特金斯，是牛津大学林肯学院研究员，英国著名化学家、化学教育家和科普作家。他已出版近70部作品，包括多部备受欢迎的化学专业教材和科普书。

生物的尽头是化学，化学的尽头是物理。本书从化学的起源、范围和组成讲起，再到原理、反应、技术、成就和未来，详细描述了化学的内涵和化学研究的清晰界限。本书通过形象的比喻，将抽象的化学概念还原为日常生活中的常见现象，让没有化学专业背景的读者也能够直观地感受到化学之美。举例来说，作者用一个碳墨打印的句号包含超百万个碳原子来展示碳原子的大小；又解释说电子云实际上是概率云，在云密集的地方容易找到电子，在云稀薄的地方不易找到电子；把电子层比喻成洋葱层；用爆炸和腐蚀来解释快反应和慢反应；用烹饪食物来帮助我们理解加热可以诱发反应；用做工程项目来形象比喻多步骤的有机分子合成；等等。本书还介绍了原子层面的四种基本反应，以及化学的分析仪器和分析技术手段，均用非常通俗的语言把讳深的原理和技术做了详细的阐述，让化学专业人士通读以后也可以大有收获。这些应该归功于作者撰写通识书本的深厚功底，引用作者的话就是："提炼出必要的知识，以及化学家在做研究的时候一般会记住或至少存留在潜意识中的要点。"本书还展示了化学取得的巨大成就，从日常生活中的衣食住行，到高科技领域的材料科学、生物医药等等，化学无处不在，无时无刻不影响着我们的生活，改变着世界。当然化学的发展，也带来了诸如环境污染、化学武器等问题，本书作者对此也毫不避讳。解铃还须系铃人，相信化学带来的一些问题，最终还需要化学家来推动科学突破、技术创新和政策改变来解决。

  正如作者所言："没有化学的生命会让我们重回石器时代。几乎所有现代世界的基础设施与舒适生活都源于化学研究。"化学家，如同自然界中的魔法师，已经创造了一个又一个化学奇迹，这些成果已经深刻影响了我们生活和生存的方式。化学，这门充满魅力的学科，正等待着每一位热爱探索、勇于创新的读者去揭开它的神秘面纱。本书作为一本化学通识书，旨在激发公众对化学的兴趣与热爱，引导大家走进化学的世界，感受化学的妙趣与力量。衷心希望有更多青年学子通过阅读本书，开启化学之旅，共创美好未来！

目 录

前　言　1

第一章　起源、范围和组成　1

第二章　原理：原子与分子　12

第三章　原理：能量与熵　26

第四章　反　应　36

第五章　技　术　51

第六章　成　就　63

第七章　未　来　81

元素周期表　90

词汇表　91

索　引　95

英文原文　103

前　言

　　我希望打开你的眼界，为你展现一个由化学构建的无与伦比的、在智识与经济层面都极其重要的世界。我必须承认，化学声誉不佳。很多人会回忆起，它是学生时代的一门科目，它难以理解、包含大量事实却难以融会贯通、伴随着难闻的气味、远离现实世界中的事件与欢乐，以至于他们不屑于谈论化学中的那些概念、名称、配制方法以及规则。之后，化学的声誉通常会变得更差，因为他们意识到了可恶的化学物质对环境造成的影响：它们逸散到野外，将灾难带给长满三叶草的柔绿色乡野草地，这里曾是色彩鲜艳的罂粟与翩翩起舞的蝴蝶的家园；将曾长有野生百里香的河岸变为条件恶劣的淤泥；让曾泛起涟漪的清澈小溪化为有毒的泥浆和令人作呕的烂泥；将风中携带的香甜气息转变为浓烈的刺鼻气味；总之，它破坏了一切事物。

　　我想改变所有类似的看法。我想让你以现代化的不带偏见的视角重新审视化学，除去那些消极的记忆与态度，替换成理解与欣赏。我想带你从一位化学家的视角观看世界，理解它的核心概念，知道一位化学家如何在为我们物质上的享受做出贡献的同时，还为人类文化做出贡献。我想解释化学家们如何思考，并解释他们所揭示的物质特性——所有形式的物质，从石头到人类——如何增添我们认识世界的乐趣。我想为你展示化学家们如何获取一种形式的物质，可能由地下抽取、挖取或从天空提取，并将其转变为另一种形式，该物质有可能为我们制成衣服、提供食物，或是让生活变得舒适。

我想与你分享化学为现代世界提供的基础设施。日常生活中，几乎没有一件物品不是由化学提供的或基于由化学创造出的材料的。如果除去化学及发挥其功用的臂膀——化学工业，就意味着除去了在建造中使用的金属及其他材料，在计算与通信中使用的半导体，在供暖、发电及运输中使用的燃料，在服装与装饰领域中使用的面料，以及在色彩缤纷的世界中使用的人造颜料。如果除去化学对农业的贡献，就意味着人会饿死，这是因为化学工业提供的化肥与农药使日益减少的耕地能够养活不断增长的人口。如果除去制药部门，就意味着人们因麻醉剂消失而痛苦，还意味着人们因药品消失而失去康复的希望。试想一个没有任何化学制品（包括纯净水）的世界：你回到了青铜时代之前，身处石器时代，那里没有金属，除了木头没有任何燃料，除了动物皮毛没有任何面料，除了草药没有任何药品，除了手指没有任何计算工具，并且只有少量的食物。

科技进步需要得到具有新颖、复杂特性的材料，要么是电、磁、光或机械性能更优良，要么只是纯度更高。保障人体健康可以减少对医院的实体基础设施及院内精密、昂贵的仪器的需求，它的进步有赖于更好、更高级的药物的研发与生产。若没有化学提供物质层面的基础结构，我们就不可能在生产、调度及节约能源方面取得进步。

更不用说，在原料与为了提高生活质量、延长生命而通过化学将原材料转化成的物质之间存在惊人差异。为实现转化，需要付出一定的代价，正是这种代价令我们感到不安，并为我们对化学产生的环境影响感到恐惧提供了正当的理由。最惨痛的代价就是，化学制造出的产品增强了我们杀戮与残害的能力，因为新型炸药及其他制剂改良之时，武器的威力也增强了。生产出的产品及在生产过程中造成的不可否认的环境影响，通常会引起更长期、民怨更大的担忧。化学通过政府选择，使社会有能力更有效地发动战争；通过商业压力，使社会有能力制造出更多的人工制品；通过个人选择，使社会更恣意浪费，从而破坏我们独特且无可替代的生态系统。

我将在本书中直面这类担忧，因为这是化学生产进步的必然结果，不仅它制造出的产品，还有它产生的废物都存在于环境中。尽管如此，重要的是记住化学的全貌，而不是单一的黑色的一面。没有化学，生活将是污秽的、野蛮的、短暂的。有了化学，生活会变得舒适、有趣、饱足，出行会变得高效，服饰会变得靓丽，生命会得到延长。我不会忽视化学黑暗与消极的一面，但与此同时我也鼓励你欣赏它富有启发性和积极的那一面。

除了以上所有贡献外，化学还有一个贡献：理解。化学通过展现事物的机制，让我们深入了解物质的核心本质。化学家可以观察一朵玫瑰，知道它为什么是红色的；可以观察一片叶子，知道它为什么是绿色的。化学家也可以观察玻璃，理解它为什么易碎；观察布料，理解它为什么柔软。当然，没有这些内在的知识,同样可以感受到自然的伟大之处,就像没有分析也可以享受音乐一样;但是在合适的时机,可以运用化学所带来的对各种形式物质的特性的深刻理解,获得更深层次的享受。我的目的就是与你分享这一深刻理解,向你展示即便只了解一点化学,也会为你的日常生活增添乐趣。

概括而言,这就是我想带你踏上的旅程。我想努力让你摆脱早年接触化学时形成的模糊的,也许不快的记忆。读完这些章节并不意味着你会获得化学学位,因为化学既有深度又有广度,既是定量的又是定性的,既微妙又浅显。即便如此,我希望你可以理解它的框架,核心概念,以及它对文化、乐趣、经济和世界做出的贡献。

最后,我想感谢帝国理工学院的教授戴维·菲利普斯,他提供了诸多有益的建议。

彼得·阿特金斯
牛津,2014 年

第一章

起源、范围和组成

贪婪。贪婪使人类踏上了一段令现今所有人都感慨万千的非凡旅程。在这里,我所说的贪婪是指对长生不老的追寻与对无尽财富的获取。得到它们的假定途径就是操纵物质来获取可以治疗病痛、实现长生不老的万灵药,还有能把类似金的东西(或是颜色相近,像尿液和沙子;或是重量相近,像铅)变成金的秘方。尽管这两个目的都未曾实现过,但是炼金术士无休无尽的试验让他们对这些物质有了相当多的了解,同时也提供了某种(常常是名副其实的)"混合肥料",使得一门真正的科学——化学——从中诞生。

天平是使得炼金术转向化学的重要仪器。精确称量物体的能力赋予了人类把数字联系到物质上的潜能。这一成就的重要性不该被忽视,因为可以把空气、水、金,以及其他各种物质与有实际意义的数字联系起来确实是一件非同寻常的事。就这样,通过附加数值,自然科学领域引入了对物质及其可发生的变化的研究(现今化学的研究范围),在其中,定性的概念可以被量化，且可以接受包括与阐述这些概念的理论的严格检验。

在物质从一种化学物质转变为另一种前后对其进行称量，从而得出了化学中所有的解释所依据的一个重要概念：**原子**。自从古希腊人在没有任何证据的情况下推测出世界上存在某种最终不可分割的微粒开始，"原子"的概念已毫无根据地在人类的意识中漂浮了两千多年。约翰·道尔顿（1766—1844）的实验证实他们的推测是有科学依据的，他通过对化学物质反应前后的质量进行分析而得出了结论：物质的基本组成要素是化学元素，它们是由不可改变的原子组成的，随着一种化学物质变成另一种，原子那种不可改变的行为方式可以通过简易的称重法记录下来。

现今，原子成了化学的货币。基本上化学中的每个解释都会涉及它们，它们或是单个存在，或是以我们所说的**分子**的形式组合在一起。原子是所有物质的组成部分：你可以看到、摸到的每样东西都由原子构成。尽管它们很小，但是说用肉眼看不到它们就完全错了。看一棵树，你在看很多原子。看一把椅子，你在看很多原子。看这页纸，你在看很多原子（即便这一页是在显示屏上显示的）。摸你自己的脸，你在触摸很多原子。摸一块面料，你在触摸很多原子。当然，单个原子是小到看不见的，但是物质是由成群的原子构成的，我们可以用肉眼看见这些成群的原子以化学物质的形式围绕着我们。然而在之后的第五章中，我会解释如今化学家们如何可以看到甚至**单个**原子的图像。

一共有一百余种不同类型的原子。至于"类型"是什么意思，我将在第二章中解释，届时我们将探究原子的内部并辨认使不同类型的原子有所区别的内部结构。每一种不同的原子类型对应着一个不同的化学元素。所以，就像有氢、碳、铁等元素一样，也有氢原子、碳原子、铁原子等，直至最近在2013年发现的 **铊**元素，它是完全无实际用途且只能极短暂存在的114号元素。（准确地说：它是116号元素，但是它前面的两个元素仍有待发现。）化学中的核心观点就是，当一种化学物质变成另一种的时候，原子自身不会发生改变：它们只会交换伙伴或形成新的排列方式。化学就像婚姻的分与合。

尽管"原子"意味着不可分割，但事实上是可以分割的。即使凭空猜想也可以得出这一结论，因为不同类型原子的存在就意味着它们拥有不同的结构，所以借由充足的想象力就可以把原子炸开并确定组成原子的所谓**亚原子粒子**。实验证实了这个猜想，我们将会在第二章中看到原子的内部结构，从而了解它们不同特性的起源。在这一方面，化学极度依赖物理学，因为物理学家阐明了原子的结构，化学家进而利用这些知识来理解由它们构成的分子和它们发生的反应。

上一句话暗示了化学的范围。它意味着需要引入物理学的概念才能理解化学。诚然，化学大量吸收了物理学家提出的诸多概念（作为回报，我们化学家为他们提供了研究用的物质）。在所有这些相互借鉴中，有两个极其重要的引入概念，其一与单个原子及其亚原子组成部分的特性相关，其二与可感知到的大的物质（例如一罐水或一块铁）相关。用更专业的话说，它们分别是**微观世界**和**宏观世界**。

量子力学是由物理学引入的一个重要概念，用于解释单个原子和分子的微观世界性质。尽管许多化学知识都是在19世纪发展起来的，但是我们并不理解为什么一些事情会发生，而另一些不会。在那时，艾萨克·牛顿的"经典力学"，即用于解释物体运动的数学演算，有至高无上的地位，因为它如此成功地解释了行星的轨道和小球的飞行轨迹，人们由此产生了这样的预期：当行星和小球被简化为原子的时候，就可以找到对化学的解释，这样牛顿力学领域也将涵盖化学。牛顿对炼金术的徒劳关注可能表明他也是如此想的。然而，在19世纪末和20世纪初，人们发现，将行星和小球简化为原子的做法让经典力学一败涂地：即便是牛顿力学中的基础概念，在被应用于原子及其组成部分的时候，也会面临土崩瓦解。这就是盲目使用外推法的危险之处。

于是，在20世纪初，1927年前后，一种新的力学诞生了，它被证实可极为成功地解释原子与亚原子粒子的特性。直至今天，在预测能力与数值精度方面，量子力学理论仍未被超越。尽管在很大程度上这一理论仍难以理解，不得不承认这是一个令人烦恼的瑕疵，但是我会在合适的时机尽力提炼出为了理解原子的行为，从而进一步理解整个化学所需要知道的知识。我们将会看到，当化学家搅拌、煮沸液体的时候，他们其实是在诱使原子按照量子力学的奇怪规则行事。

另一个由物理学引入的重要概念是**热力学**，它用于解释宏观世界中大的物质的特性。热力学是研究能量及其转化的科学。它的出现的一个很重要的原因在于维多利亚时代对蒸汽机的依赖，它推动社会在字面意义上前进也在经济方面进步，但它很快就成了构成化学的一个重要部分。我们这门学科的物质基础是原子，但它们所发生的变化受制于能量的控制与驱动。我们将会看到，能量不仅在燃料燃烧的时候被释放出来（这是能量参与化学反应的一个明显、有用却原始的方式),它还支配着原子的一般行为,例如原子可以形成怎样的结构、原子可以发生怎样的组织变化,以及这些变化可以以怎样的速率发生。能量还会以一种微妙的方式成为化学反应的推动力,因为反应是在能量的驱动下进行的,我将会在第三章中对此进行详述。因为能量与化学的结构密切相关,所以尽管热力学发源于工程学,它却在化学中起着很重要的作用,这就不足为奇了。

化学向下深入物理学寻求解释(还通过物理学向下深入数学寻求量化表述),向上延伸至生物学实现不计其数的非凡应用。这并不意外,因为生物学只不过是化学的延伸。这种论调可能类似于宣称社会学是粒子物理学的延伸,在生物学家对此表示不满之前,请允许我详加解释。有机体由原子与分子构成,这些结构可以用化学解释。有机体发挥功能,即存活,是通过体内复杂的反应网络进行的,这些反应可以用化学解释。有机体利用分子结构与化学反应进行繁殖,这些同样都是化学的一部分。有机体通过改变分子结构的方式,比如通过嗅觉和视觉,对它们所处的环境做出反应,所以这些反应(我们所拥有的全部五种感官)都是化学的延伸。即使像物种进化与物种起源这类超宏观现象,也可当作热力学第二定律鬼斧神工般的作品,因此也算化学的一个方面。一些有机体可以思考世界的本质,我主要指人类,而作为这些思考的基础和表现形式的心理过程正是来源于复杂的化学反应网络。所以,生物学确是化学的延伸。不管我真正的想法如何,我并不想将这种看法强加给他人——一切令生物学家感兴趣的东西,例如动物的一般行为,也只是化学的延伸,我限于认为所有有机体的结构、反应以及生理过程都是化学的。因此,化学渗透生物学领域,为我们对有机体的理解做出了不可估量的贡献。

我们人类是社会意义上精巧的有机体,可以建造东西。我们可以制造人工制品。我们可以开采地下的矿石、抽取深海中的液体,还可以收集天空中的气体,并致力于将所有这些原材料转变为我们需要的任何东西。化学的一个方面就是将这些原材料转变成可铸造、锻造、纺织、黏合、食用或仅可燃烧的物质。尽管化学家们可以靠边站,让铸造工来铸造、锻造工来锻打、塑形工来塑形,总的来说,就是让制造工来制造,从而创造出最终的人工制品,但正是化学家们提供了原材料,提供了现代科技社会的基础设施,从而为世界经济及个人与国家的风貌做出了巨大贡献。

就像我在前言中强调过的那样,在这所有光辉之下也存在黑斑与污点。化学诚然赋予了人类残害与杀戮的能力,在这本关于化学是什么的小书中,我们不应无视它供应的炸药、神经毒气以及它对我们脆弱的环境无心或故意地施加的压力。我会在后面的内容中直面这些问题,但在此——为了强调个人评判的重要性——我想请你暂且忽略化学对现代世界的所有贡献,这会将你带回痛苦的、危险的、不舒适的、精神受压抑的石器时代,然后请你思考此时此刻的黑暗是否遮掩了光明。

化学的分支

化学的范围如此广,如若不设置某种框架,那么我对它的介绍以及这门学科本身就会像一头没有脊椎的鲸鱼一样没有方向地翻来滚去。化学家们摸索到了一种框架,便于他们开展研究工作，组建志趣相投的团体，还可以像独自制定政策、发展经济的国家那样制定他们的程序。有别于大多数国家，这里的边界是模糊的，通常在两个领域的重叠处会取得惊人的进展。尤其是当学科像目前的化学一样成熟，每个活动领域都被仔细探索过时，灵感可能会如同在艺术领域内一样，出现在肥沃的重叠边界上以及化学与其他学科重叠的前沿领域中，从而取得丰硕的成果。

为了我们的目的，即在本书中了解化学的总体结构，我们需要了解化学的不同分支并大致了解它们所关注的问题。化学的这些分支仍遍布于大学的科系、课程，以及报道新发现的期刊中，所以对它们的介绍仍是化学入门手册的重要组成部分。但请留意：不管是智识的还是科系的，边界正在消解。

化学最广泛、最重要、最传统的，如今仍被广泛使用的分类方法是分为物理化学、有机化学和无机化学三个分支。

物理化学处于物理学和化学的交界处（它的名字由此而来），它研究化学原理，就像我们之前提到的，它主要包括解释原子、分子结构的量子力学和评估能量的作用、利用的热力学。它还关注宏观和微观层面上反应发生的速率。在后一研究中，物理化学追踪单个分子在反应过程中分开、随后重组成不同化学物质的亲密生活。物理化学的一项主要活动就是解释研究方法，尤其是"光谱学"。

我们将在第五章中了解到，光谱学使用各种光将分子内部的信息呈现到观测者眼中，越来越多的是合成眼中。为了解释由这些极其复杂的方法获得的数据，物理化学家必须拿出他们的所有法宝，尤其是量子力学。事实上，在这一领域内，化学家和物理学家的工作没有明确的界限，以至于对某些使用接近物理学家的方式研究单个分子行为的科学家而言,物理化学这一名称常被称为**化学物理学**。

有机化学是有关碳化合物的化学。一个元素可支配一整个化学分支的事实证实了,看似普通的碳却本领非凡。元素周期表是化学家使用的有关元素化学性质的图表,碳处于元素周期表的中间点,在很大程度上,它并不会选择与其他元素形成化学联系。值得一提的是,它满足于与自身结合。鉴于碳这一不温不火的特性,它可以形成复杂得令人惊叹的链与环。有机体之所以具有生命,正是因为有了这令人惊叹的复杂性,所以碳化合物是生命的结构与反应方面的基础设施。如今数百万的碳化合物涉及极其广泛的领域,以至于不出意料地,为了能对它们进行研究而发展出了一整个化学分支,并发展出了特殊的研究方法、命名法体系以及看法。

为什么被称为"有机"?这是因为碳元素造就了分子的复杂性(除了几个例外,比如简单的二氧化碳分子),以至于我们起初认为只有大自然才能形成它们。换而言之,依据"活力论者"的观点,它们是有机体的产物。活力论的终结起始于1828年,当时有研究证实,一种简单的无机矿物可转化为具有"有机"特征的化合物(即尿素)。即便争论激烈地持续了一段时间,但从那时起,有机化学中的"有机"就成了一个古语;然而方便的古语难以替代,于是这个词语存留下来,但现在仅用于表示"碳的化合物"。

除了碳元素外,还有百余种其他元素。对它们的研究属于**无机化学**的范畴。可以想到,作为一个涵括一百多种特性截然不同的元素的学科分支,无机化学是一个重要且庞杂的研究领域。通过对该学科进行各种细分，可以在某种程度上约束它的杂乱性。一个重要的分支就是**固体化学**，其研究对象为无机固体，例如超导材料与使计算普及成为可能的半导体材料。很难不把无机化学类比于由百余件乐器组成的管弦乐团，担任指挥兼作曲家的化学家适当地排布乐器，从而演奏出以不同方式组合而成的交响乐。

碳也没能逃过无机化学家那双元素周期表扫描眼。一些较简单的碳化合物，像是我之前提到过的二氧化碳，以及致命的毒气一氧化碳、构成景观的白垩和石灰石，有机化学家对它们并不感兴趣，乐意将它们排除在研究范围之外，它们按照惯例被归为无机物。然而，就在这两个分支交汇之处，有一种化合物既是碳原子的复杂集合体，又含有各种金属原子。很多此类化合物是化学工业中重要的催化剂；还有一些对有机体的机能起到了重要作用。这就是**金属有机化学**这一跨学科领域，它是有机化学家与无机化学家之间卓有成效之合作的成果的最佳体现。

化学与其他学科的重叠

以上是化学的三个主要分支。这份清单绝对还没有穷尽化学家为了更好理解这一学科而划分它的方法，然而所有其他学科的科学家都会以不同比例从这三个分支吸取技术、概念及灵感，再在这一混合物中加入其他学科的内容。罗列出所有的分类方式将会是一项艰巨的任务，即便如此，还是应该知道其中最常见的一些。

分析化学是一项古老探索的现代延续：查明物质中有什么。矿石里有什么？金还是铪？原油里面有什么？除了未加工过的碳氢化合物还有什么,具体是哪种碳氢化合物?你制成的化合物是什么?你能推测出其原子排布方式吗?这些都是分析化学家努力回答的问题。尽管试管、锥形瓶和曲颈瓶仍会出现在他们的方法中,但是现在他们大部分的研究都会在精密仪器中进行,一些仪器基于光谱学,而另一些则使用由无机化学家和物理化学家发明的方法。我会在第四章进一步探讨这些方法。**法医化学**起源于分析化学,其中分析化学的方法被用于法律目的,可以确定或排除嫌疑人,以及分析犯罪现场。

生物化学是有机化学对生物学的回赠,有时还混入一些无机化学的成分。它专注于研究构成生命的结构与反应,分析将食物转化为行为(包括仅限于大脑的行为:思考)的代谢途径。有机体仍是一个极其重要的有机分子宝库,因为大自然已经用了几十亿年的时间去探索合适的结构,而生物化学家在发现有机体内的有机分子与解答这些有机分子如何在体内的工蜂(我们称之为酶的蛋白质)的调控下被制造出来方面发挥了重要的作用。谈及物种灭绝,有一个虽然以人类为中心但又极为重要的担忧:物种灭绝会消除经过几百万年才出现的复杂分子的起源信息。

工业化学这个名字不言自明。它是化学家与工程师的碰面,试管及类似容器中的反应被放大到巨大的规模,从而适于商业化。工业化学家们对经济与国家间的贸易做出了巨大贡献。仅在英国,化学品就贡献了20%的国内生产总值,而在美国,所有制成品中超过96%都直接与化学相关。这些与化学制品相关的数字绝对不容小觑。现今,工业化学的一个主要关注点是**绿色化学**,其目的在于减少浪费,从而提高经济性,并把对环境造成的影响降到最低限度，由此增强可接受性与可持续性。

化学对其他学科的贡献

尽管在知识领域内，化学对周围学科有诸多亏欠，但这些学科也对化学有所亏欠。

物理学对化学有所亏欠，尤其是在电子学和光子学领域内（用光取代电子来传递信息和处理数据）。化学家们创造了半导体，没有它，计算将会仅限于最早出现时的工业规模。他们也制成了用于光导纤维的玻璃，没有这项发明，信息的传送将寸步难行。

生物学对化学亏欠极大，尤其在**分子生物学**出现后，它的诞生源自对DNA结构的确定以及认知到它是基因信息代代相传过程中的载体。几乎可以毫不夸张地说，在生物学的主要特点，繁殖的化学成分被确认后，它就成为自然科学的一部分。分子生物学实际上是化学的一个版本，如今化学的成熟已赋予生物学前所未有的活力。生物学与化学的合作，也就是我们所说的**药物化学**，是化学对社会的重大的、毋庸置疑地令人满意的贡献之一。

社会对化学还有另一样巨大的亏欠，就像我在前言中所提及过的，化学在医药、农业、通信、运输，以及各种形式的建筑、制造和装饰等领域的物质贡献无处不在。我们每个人也对化学有所亏欠，我在前言中也提到过，它赋予了我们欣赏世界的第三只眼。

这一切都源于对化学的理解，我现在将开始细述。

第二章

原理：原子与分子

元素周期表是任何化学讨论的核心，这一结构编排方面的杰作主要由德米特里·门捷列夫（1834—1907）在19世纪制定，在20世纪我们可以解释原子结构之后，其依据才被人们理解。元素周期表的无处不在证实了它的重要性：它悬挂在实验室和教室的墙上，印在每本化学入门教科书上。在本书的最后就附有其中的一个版本。不过，它的重要性不该被夸大。做研究时，化学家们不会为了获取灵感而每天清晨注视它或在白天频繁参考它。他们当然会将它装进脑中，因为它的重要性在于总结了不同元素之间的关系，并在排列元素相关信息时起到关键作用。也许它最重要的用途在于化学教学，有了这张表，学生们就不需要面对学习一百多种元素的不同性质这项令人发怵的任务了，而是可以通过元素在表中的位置推断出它们的性质，并识别和轻松记住性质的变化趋势。事实上，门捷列夫正是在准备编写化学入门教科书的时候制定了元素周期表。

元素周期表描述了物质的一个非凡特点：元素间是相互关

联的。如今，我们对元素周期表过于熟悉，以至于很容易忽略这一特点。但设想你自己身处这张表制定之前的年代。那时，你就知道名为氧气的气体与名为硫的黄色固体，但几乎肯定从未想象过它们之间会有任何关系。你还知道在绝大多数情况下为惰性的氮气与燃烧时发出白光的固体磷，但不会想到它们之间有任何关联。那么红色的铜、闪亮的银，以及熠熠生辉的金，你会想到它们属于同一族吗？肯定不会！不同形式的物质究竟如何才可以成为兄弟？你甚至不会想到不同物质间的族系关系这一概念。

不过，元素周期表揭示了元素间确实相互关联。氧和硫是表兄弟，在表中处于相邻位置；氮和磷也有类似关系；铜、银和金是同族成员，处于同一列。它们极不相同的外观只是表面上的差异，因为当研究它们参与的反应与它们形成的分子的时候，就会发现这些亲戚之间有深刻的相似性。这些相似性源于它们的原子结构，所以为了理解这些相似性，我们现在必须将目光转向这些原子。

原子的结构

要理解原子的结构，就必须了解量子力学和它所有艰深晦涩的知识，我在第一章中提到这一点时可能稍微有一些不安。但是，我同样提到，我会从这一非凡理论中只提炼出我们需要了解的概念与信息。脑中设定了这一限制后，我们就会发现原子有相当简单的结构，也很容易理解元素之间的关联，沿着这一思路，我们还会理解为什么一些原子的组合是可以的，而其他的则不可行。

原子的基本结构包括一个由电子云环绕的原子核。这就是"核式原子",是由欧内斯特·卢瑟福(1871—1937)在1911年首次提出的一种原子模型。原子核带正电,电子带负电,正是这些相反电荷间的吸引使原子得以存在与维持。众所周知,原子非常小:在这句话末的句号(印刷版本)中就包含了超过一百万个碳原子。原子核甚至更小:如果将原子放大到足球场大小,那么原子核就类似于场地正中的一只苍蝇。

我会从原子的中心开始解释,然后转向外层。原子核由两种亚原子粒子组成:质子和中子。就像它们名称中 p 和 n 所表明的,质子(proton)带正电(positive),中子(nerton)则呈电中性(neutral)。除此之外,它们极为相似,质量几乎相等。在原子核中,它们紧密地结合在一起,需要尽很大努力——例如核爆——才可以将它们分开。在大部分化学反应中,原子核只释放出相对微弱的能量,它保持不变,只被动地参与试管、烧杯和烧瓶中的反应,但这一参与又十分重要。

原子核内的质子数决定了原子的化学身份。氢原子有一个质子,氦原子有两个,碳原子有六个,氮原子有七个,氧原子有八个,以此类推,直到铊原子,它有116个质子。原子核内的质子数被称为元素的**原子序数**。这样,我们发现了元素第一个非同寻常的特性:可以根据原子序数排列。元素不再是随机而杂乱的。它们按照一定的顺序排列:氢、氦……铊。除此之外,因为原子序数可用作一种点名册,所以化学家们和物理学家们知道,除了113号与115号元素外,他们已经确定了116号以内每个原子序数的元素(截至2014年)。他们知道,除了这两个元素及116号以外的元素,其他元素均已被发现。

中子不过是这场点名的过客。原子核中的中子数与质子数大致相等，可略有变化。因为中子数不会影响原子序数，所以同一元素的原子可以有数目略为不同的中子，因而质量也略为不同。同一元素的不同原子被称作**同位素**（isotopes），因为它们在元素周期表中处于相同的位置（isos ="相同的", topos ="位置"）。因而，氢元素有三个同位素：氕（一个质子，无中子）、氘（一个质子，一个中子）和氚（一个质子，两个中子）。至今为止，氢元素的同位素中丰度最高的是氕；氚的原子核很难结合在一起，且具有放射性，几年后衰变时会产生辐射（它的半衰期是12.3年）。氘原子是"重氢"，它的每个原子质量大约是普通氢原子的两倍。它可以与氧原子结合形成"重水"，因为氘原子比普通氢原子重，所以重水大约比普通的水重10%。

原子序数、质子数，即原子核所带的正电荷数，决定了原子核周围的电子数。一个**电子**携带的电荷数与一个质子相等，但是性质相反。所以，原子呈电中性的前提为，原子核外的电子数一定等于原子核内的质子数。换言之，电子数等于原子序数。所以，氢原子（原子序数为1）有一个电子，碳原子（原子序数为6）有六个电子，以此类推，直到鿫原子，它有116个电子。电子比质子和中子轻得多（相差将近2 000倍），所以它们的存在几乎不会影响一个原子的质量。它们对元素的化学和物理性质有重大影响，几乎所有化学反应都可以归结于它们的行为。

化学家对原子核没有太大兴趣，除了用于确定环绕于其周围的电子的数量。但有一个例外，就是氕原子核，它是单个质子。我将会在第四章中解释它的特殊之处。

就像我之前提到的，所有化学反应中，原子核都保持不变。

换言之，化学反应不改变元素的身份。一下子，我们就可以理解为什么炼金术士们迫切想找到把铅（82号元素，原子核内有82个质子）变成金（79号元素，原子核内有79个质子）的方法，却注定徒劳无功了：沮丧地加热、搅拌、锤击、冲压并不能把其中紧密结合的三个质子从原子核中提取出来，而这是"核嬗变"，即从一种元素到另一种元素的转变所需要的。核嬗变可以发生，但这是**核**反应的结果，属于核能和核物理学的范畴。化学家在处理核反应的后果方面起着重要作用，尤其在制备核燃料与处理核废料方面，但是在**化学**反应中所有原子核都保持不变，现阶段我们只考虑化学反应。

原子中的电子

我现在要开始讨论环绕原子核的电子云，它们有极为重要的性质。我需要更精确地解释这些云的性质与结构，因为它们不仅仅是一个旋动的薄雾似的区域。

电子层层环绕原子核，但不像真正的云那样层层堆叠，而是每一层都环绕着整个原子。我要简单地解释一下一个电子如何呈现为一片"云"。这种云实际上是概率云：在云密集的地方，容易找到电子；在云稀薄的地方，不易找到电子。

量子力学的定律规定：环绕原子核的最内层最多容纳两个电子，第二层增加至八个电子，第三层增加至18个电子。我们无须了解更多，但是类似的变化模式随着电子数增加而无限地延续。这种模式意味着，在氢原子中只有一个电子环绕着原子核。碳原子中的六个电子，其中两个形成最内层的电子云，另外四个在稍外一层形成环绕的电子云。你可以把原子想象成洋葱层一样的电子云环绕着原子核，先填满一层再开始填充下一层。我们不需要知道为什么这些连续的电子层具有特征性的电子数（2、8、18……），只有从量子力学层面才可以完全理解这一点。

现在，你们将了解元素周期表的构造与元素间的家族关系。请看本书末尾的元素周期表，从有单电子的氢元素开始，再到有两个电子的氦。这样第一层就填满了，同时我们来到了周期表的最右边。锂所需要的下一个电子就要成为下一层中的云。随着电子数增加，横跨这张表，经过碳、氮、氧，直至氖元素，这样就填满了这一层，氖是类似于氦的一种气体。下一个增加的电子需要填充至下一层云中，如此，我们到了表最左边的钠，它与上一行中的锂极为相似，在它们已被填满的云核之外都有一个电子。

现在一切应该都很清楚了：周期表的排布代表着云层的填充，在表的最左侧，最外层只有一个电子，在表的最右侧，最外层已经填满。云层填充的顺序在前两行之后变得有些混乱，尽管行的长度是我们已见过的数字，也就是2、8、18……并且可以勉强辨认出来，可它们是以奇怪但可被理解的方式排列的（元素周期表的排列模式为2、8、8、18、18……）。这种现象在技术上完全可以理解，在此不做过多说明。

一个要点在于，同一列内的元素有极为相似的电子云排布模式。这就是家族关系的起源：氧与同族内位于下一行的硫具有相似的电子云模式，唯一的区别在于硫最外层的六个电子比氧最外层的六个电子处于更高的一层中。类似地，磷最外层的五个电子比上一行中的氮最外层的五个电子处于更高的一层中。

人们常说，原子大部分地方是空的。这并不是真实情况。

云状分布的电子充满了苍蝇大小的原子核周围的体育场大小的整个空间。诚然，某些地方的云层很薄，但它确实存在且无所不在。原子大部分地方是空的，这一说法源自一个过时的观点：电子就像微小的点状行星，远远地围绕着原子核快速旋转，其间空隙很大。量子力学以我提过的云状分布取代了这一图景，尽管云层的某些地方极为稀薄，但它仍充满整个空间。

原子怎样成键

化学的主要关注点并不在于单个原子，而在于这些原子以各种各样的方式发生联系而形成的化合物。我们已发现了几百万种此类联系，还有更多我们知道存在但还没有发现和命名的联系。我们身边环境的丰富性正是由于大量化合物的存在，化学家花费了大量时间构建新的原子组合或将化合物分解以了解它们构建的方式。为了高效地进行这项研究，他们需要了解原子如何结合在一起以及什么因素控制了它们的结合，即**化学键**。

是什么让原子结合在一起形成可辨识的化合物，如水、盐、甲烷和DNA？原子间的结合方式是随意的，还是受无论化学家们如何尝试，都无法避开的自然法则限制？为什么物质世界中存在多样性，却显然不是随机的多样性？这些问题可以反过来问：为什么宇宙中的所有原子不聚成一块巨大的固体？

所有这些问题的答案都在那些云层之中。一般来说，获得一整层电子云的原子具有能量方面的优势。有很多种形成一整层电子云的方法。一种是从最外层失去电子。如果最外层电子不多，原子就有可能这样做，这意味着元素周期表左侧元素的原子更可能使用这种方法，也就是在每一行、每一个新云层开始之处。另一种情况是，如果原子最外层已经有许多电子，那它可能从其他地方获得电子，从而填满这一层。如果最外层几近满了，获得电子的可能性就会很高，在元素周期表右侧、每一行右端的元素的原子通常会使用这种方法。还有另外一种填满电子层的方法：原子们可以共用最外层的电子。这种情况会发生在一个原子不愿完全失去一个电子的时候，因为这样做没有能量方面的优势。碳正是以这种中庸的方式形成了大部分非凡的联系。

就像前面提到的，原子呈电中性，其中所有电子携带的负电荷总量可以匹配并抵消掉原子核内所有质子携带的正电荷总量。当原子获得或者失去一个电子的时候，电荷的平衡会被打破，原子就变成了**离子**。离子就是带电的原子；如此命名是因为它会在电场的作用下移动，"离子"（ion）在希腊语中就是移动的意思。获得一个或多个电子的原子带负电，被称为**阴离子**（anion）。失去一个或多个电子的原子带正电，被称为**阳离子**（cation）。前缀 an 和 cat 源于希腊语中的"上"和"下"，反映了一个事实：带相反电荷的离子在电场的作用下会向相反的方向移动。我用一句话总结上一段中的话：位于元素周期表左侧的元素很容易失去它们最外层为数不多的电子，成为阳离子；位于表中右侧的、最外层电子云几乎已填满的元素很容易获得电子，成为阴离子。

我们发现了一个很重要的成键机制：由于相反电荷相互吸引，而阴、阳离子带有相反电荷，所以形成阴、阳离子的不同原子会结合在一起形成化合物。常见的食盐，氯化钠就是一个形成这类化合物的绝佳例子。钠（Na，源于拉丁文名字 natrium）位于表的左侧，它极其容易失去最外层的单个电子，成为阳性的钠离子，写作 Na^+。氯（Cl）位于表的右侧，很容易获得一个电子，从而填满最外层，成为阴性的氯离子，Cl^-。（注意名称的微小变化，从氯变成氯化物。）这些离子聚集在一起，形成氯化钠，一种由离子间的相互吸引作用而聚成的一块坚硬固体。我强调过原子非常小，即使是一块微小的物质样品也包含大量的原子。在离子层面上，你就是群星之间的擎天巨神阿特拉斯，因为当你拿起一小块盐的时候，你手中的离子数量要远多于可观测宇宙中的星星。

至此，你应该可以理解为什么从某地开采，或是从某片海里提取的盐与从世界另一端开采或提取的样品有同样的成分了。钠原子的最外层有一个电子，氯原子的最外层有一个电子空位；所以一个钠原子与一个氯原子唯一可能的组合方式就是通过失去与获取电子形成离子，从而结合在一起。食盐的主要成分是 NaCl，其中钠离子与氯离子的比例为1∶1。像 Na_2Cl（离子存在的比例是2∶1）或是 Na_2Cl_3（离子存在的比例是2∶3）等这类化合物绝不会存在。现在我们应该逐渐清晰地认识到，大自然对于哪些联系是可以形成的、哪些是不可以形成的有严格的法则。

我至今所描述的成键方式被称为**离子键合**。这类成键方式一般会形成坚硬、易碎的固体，只在高温的条件下熔化。地貌中的花岗岩和石灰岩就是通过离子键的方式将原子结合起来组成材料的例子。我们站在这些岩石上而不沉陷的原因可追溯到一个事实：围绕着那些原子核——那些原子现在呈现为离子——的电子层是完全填满的，我们自身原子中的电子云不能占据那些原子中电子云已占据的相同位置。我们的骨头大部分也是通过离子键形成的，从而为我们的器官提供了一个足够坚固的框架。

我们柔软的器官、皮肤，以及覆盖我们皮肤的布料、像布料一样覆盖石灰岩的植被、景观中的软装饰，都显然与通过离子键形成的物质有截然不同的特征。尽管离子可能会存在于其中，但它们无关这些构造的主要特征。在此涉及的成键方式为原子以**共用**电子的方式填满它们的电子层，从而结合在一起。这种成键方式被称为**共价键合**（covalent bonding），co 意为合作，valent 源于拉丁文中的"力量"一词：valete! 是罗马语中"再见! 要坚强！"的意思。

一个共价键的简单例子就是水分子的结构，几乎所有人都知道水分子是 H_2O。氧的最外层有六个电子，可以接纳两个电子从而填满最外层（记住，这一层最多可以容纳八个电子）。氢原子有一个电子，再得到一个电子（记住，氢的第一层，也是最内层仅可以容纳两个电子）就可以填满最外层（它仅有的一层）。只要两个氢原子愿意与氧共用两个电子，就可以完成共用：每个氢原子通过共用得到两个电子，氧原子通过共用得到八个电子。我们立即就可以理解水不可能是 H_3O 或是 HO_2：H_2O 是唯一能填满所有原子最外层的成键模式。氨，NH_3（N 表示氮）也是同样的道理，因为氮原子最外层有五个电子，还需要三个电子来填满这一层。三个氢原子每个都愿意与氮共用一个电子，这就满足了条件。甲烷，CH_4 也以同样的方式成键，因为碳有四个电子的空位。

你需要像化学家一样意识到离子键和共价键之间有一个很重要的区别。离子键合会产生巨大的离子集合体：本质上就是成块的物质。共价键合一般会产生离散的原子集合体，比如 H_2O。换言之，共价键合会产生独立的分子。这一区别极为重要，一定要牢记。因此，所有气体都是由分子构成的，比如氧气

（O_2分子）和二氧化碳（CO_2分子）；绝不会有通过离子键形成的气体。即便这种气体可以形成，所有的离子也会立即聚集形成一块固体。绝大多数常温下呈液态的物质都是由分子构成的，这是因为分子要可以移动而不被其邻居的强大吸引力困在某处。水就是一个很明显的例子，汽油也是。

共价键合也可以产生固体，所以你不能推断每种固体都是以离子方式成键的：所有离子化合物在常温下都是固体，但并不是所有固体都是离子化合物。蔗糖就是一个以共价方式成键的固体化合物的例子，它是由碳、氧、氢组成的共价化合物，分子式为$C_{12}H_{22}O_{11}$，每一个蔗糖分子中的原子都由共价键连接在一起，形成一个错综复杂的网。

共价键得以形成的一个重要因素就是，电子对的绝对重要性。20世纪最伟大的化学家吉尔伯特·路易斯（1875—1946）指出了这一重要性，但量子力学仍未对此给出解释。对我们而言，每一对共用电子算作一个共价键，所以我们只要数出原子共用的电子对的个数，就可以轻而易举地数出它们形成的键的个数。一个共用电子对被称为"单键"（以—表示），两个原子间的两个共用电子对被称为"双键"（以═表示），三个共用电子对被称为"三键"（以≡表示）。只有在极少数的情况下才可以形成更多的键，所以我们只需要知道这三种共用电子的方式。H_2O中的每个氢原子都以单键的方式与氧原子结合。二氧化碳分子有两组双键，可表示为O═C═O。三键更为罕见，我不会进一步谈论它们，此处只举出一个例子，即在氧-乙炔焊中使用的气体乙炔，它表示为H—C≡C—H。

这一描述背后潜在的一个问题就是，为什么两个电子（即一个"电子对")对共价键的形成如此重要。问题的答案深藏于量子力学之中。提示之一就是，所有的电子都会自旋。如果两个电子沿相反的方向旋转，从而锁定它们的自旋，它们就可以达到一个更低的能量状态。自旋锁定的重要性的另一个表现就是，我们之前提到过，每个云层中的电子都是偶数个（2、8，等等）。在法语中，未配对电子被称为"electron célibataire"①，这可能就是法国人对配对重要性的暗示。

金 属

到目前为止，我隐瞒了第三种化学键的存在。大部分元素都是金属：比如铁、铝、铜、银和金，我们会看到，金属在化学中起着很特殊的作用。一块金属是由原子组成的板，但是这些原子是通过离子键还是共价键结合在一起的呢？我们很快就会碰到一个难题。在这块板中，所有原子都是相同的，所以不太可能一半形成阳离子、另一半形成阴离子，这样就可以排除离子键合的方式。如果所有的原子都以共价的方式成键，我们应该会看到一个刚硬的固体（例如钻石，其中的碳原子就是以这种方式结合的）；但是金属可以被锤打成不同的形状（它们具有"可塑性"），可以被拉成金属丝（它们具有"延展性"）。它们也具有光泽（可反射光）且能传导电流，即电子的流动。

金属原子通过**金属键合**结合在一起。这不仅仅是同义反复。有关其特性的线索在于一个事实，我们可以看到，所有金属都位于元素周期表的左侧，这些元素的原子最外层只有几个电

---

① célibataire 意为未婚的、独身的。——译注

子，极易失去。可以这样设想金属键合，想象所有这些最外层的电子都脱离其母原子并且聚集在弥漫整个原子板的海洋之中。剩余的阳离子位于这片海洋之中，并积极地与之相互作用。如此，所有的阳离子都结合在一起，形成了一块坚实的固体。这块固体具有可塑性，因为锤子敲打这些阳离子的时候，它可以像一片真正的海洋一样快速地应对其中阳离子的位置变化。电子也可以迅速回应阳离子位置的改变，让金属拉伸成丝。这片海洋中的电子没有被固定在特定原子的周围，它们可以流动，可以在电场的作用下在固体中移动。金属之所以具有光泽，是因为这片海洋中的电子可以回应入射光线的电场所引起的振荡，海洋的振荡随之产生我们可感知到的反射光线。我们注视镜子表面的金属涂层的时候，看的是金属中电子海洋的波浪。

在这个阶段，我们学习的化学知识是，在自然状态下是金属的那些元素很容易失去它们最外层的电子。所以，在形成阴离子的前体存在并且可以接受这些被遗弃的电子时，这些金属元素的原子也可以形成阳离子。位于元素周期表最右侧的元素是电子的接受者，因为它们的最外层有一两个空位，可以容纳外来的电子，即那些形成阳离子的原子给出的电子。所以，离子化合物（请记住氯化钠的例子）通常是由一个周期表左侧的金属元素与一个周期表右侧的非金属元素形成的。

只要记住上述总结，你就会开始像化学家那样思考，能够预测元素组合形成的是哪种化合物，还可以预测它会有什么性质。你也会开始了解元素周期表与元素及元素形成的化合物的性质有何联系，还会开始了解邻近元素之间的家族关系在实践中如何体现，这种关系源自云状的电子和周期性重复的相似排布。

章末结语与下一章

　　以上就是化学结构方面的核心原理。这些内容可以归结为原子存在、对原子结构的认知,以及电子的行为。我们接下来将会谈论促进与限制化学反应的"胡萝卜和小拖车":能量。

第三章

原理：能量与熵

原子是理解化学的一条大河，另一条河由能量构成。为了理解化学反应为什么发生、怎样发生，还有各种化学键为什么形成、怎样形成，化学家们会思考这些过程发生时的能量变化。化学家也对能量本身感兴趣，例如燃料燃烧或食物——一种生物燃料——在有机体内得到利用。就像我在第一章中提到的，对能量及它可以发生的变化的研究属于热力学的范畴，我们现在就要转向这一话题。

我已经写过很多有关热力学定律的内容了，就不在此赘述了。就像我在第二章中对量子力学所做的一样，我会提炼出必要的知识，以及化学家在做研究的时候一般会记住或至少存留在潜意识中的要点。

一些热力学知识

化学热力学的精髓在于其需要记住的两点：能量的数量与能量的品质。热力学**第一定律**表明，宇宙中的总能量是恒定且不能改变的。能量可以以不同方式分配，可以由一种形式转化为另一种，但是任何过程都不能改变其总量。所以，第一定律为能量的变化设定了界限：任何变化都不能改变宇宙中的能量总量。热力学**第二定律**表明，任何自然变化都会导致能量品质的降低。这一定律可以用**熵**这一术语来进行更专业的表述，熵是一个能量品质的指标，在某种意义上，熵越高能量品质越低，这可以表述为"宇宙中的熵总是在增加"。在"混乱度"的精确意义上，熵是混乱度的指标，更混乱意味着更高的熵。第二定律可以被当作对包括化学反应在内的自然变化推动力的总结，因为只有让宇宙中总能量的品质降低的反应才能自然发生。简言之，混乱度增加，则结果变得更糟。因此，热力学的精髓可以总结为：第一定律明确了在一切可能发生的变化之中实际可行的变化（总能量不可改变），第二定律明确了在这些可行的变化之中自然的变化（熵必须增加）。

能量的角色

化学家以各种方式运用这两个概念。在传统的观念中，他们认为：如果原子间的重新组合可以降低能量，那么在化学反应中化学键就会形成或被取代。尽管这一观点对像我这样挑剔的人来说错得离谱，但它像很多错误的说法一样，是一个方便且易于记忆的经验法则。它错的原因在于违反了第一定律：总能量是不可改变的。正确的解释应为，如果一个过程，例如成键，向周围释放能量，那么这就代表着能量的退降，因为它会扩散且很难再被利用：能量释放增加了宇宙中的熵，所以是一个自然的过程。"降低能量"这一经验法则在大多数情况下成立，原因在于：

以这种方式释放能量会导致熵的增加。工作中的化学家经常理智地使用这一经验法则,我亦将如此。然而,在使用它的时候我仍然会交叉手指,就像伽利略想到地球围绕太阳旋转时低声说"eppur si muove"("它在转动")那样,我也会在心底对自己说,是熵在增加而不是能量在降低。

如果原子间成键可以导致能量降低(交叉手指),化学键就会形成。形成离子键(离子间的吸引)还是共价键(共用电子对)取决于,电子从一个原子转移到另一个原子而形成离子的这一过程释放的能量更多,还是部分转移、共用电子这一过程释放的能量更多。所以,两个元素间形成离子键还是共价键,可以通过考虑不同成键方式的能量变化来判断。

元素的特征化合价也是如此,化合价是指元素可以形成的共价键的个数。化合价是一个元素的化学性质及与邻近元素的家族关系的另一方面,可以从它在元素周期表中的位置推断得出。我们在第二章中看到,最外层有两个空位的氧原子可以与两个氢原子达成一致,形成 H_2O,以此方式填满这一层,H_2O 可更准确地表示为 H—O—H,这表明氧的化合价为2。结合更多的氢原子需要电子占有一个远离原子核的新云层,这么做并没有能量方面的优势。形成更少的化学键并不会优于形成两个化学键。所以,从能量角度而言,氧的化合价为2。在第二章中,我们还看到了氧化合价为2的其他例子,即氧与碳结合形成二氧化碳(CO_2),可准确地表示为 O═C═O。从这个例子同时可以看出,碳表现出了特征化合价4,与甲烷(CH_4)中碳的化合价一致。

我们现在可以看出,如何从元素周期表中元素的位置推测出它的特征化合价:碳的特征化合价是4,与之相邻的氮是3,与氮相邻的氧是2。同样的规律也适用于下一行中的相邻元素：硅的特征化合价是4，磷是3，硫是2。我们又一次看到，能量方面的考量如何与原子结构的概念结合在一起——尤其在填充原子云状的层结构方面，从而解释相邻元素的相似性。

了解反应中的能量

很多化学反应都会释放能量，例如天然气或汽油的燃烧。这一过程不只是化学键形成的时候释放能量，因为在起始原料，比如甲烷中，已经有原子成键结合在一起。在很多反应中，化学键一定要断裂从而形成新的化学键，在此我重点讨论燃烧。释放出的能量应为这两步中能量变化的差值。例如，在甲烷燃烧的过程中，由于它与氧气（O_2）的反应，甲烷中的四个碳氢键与氧气中连接两个氧原子的化学键都必须断开，这一步会消耗大量能量，此后，二氧化碳中新的碳氧键与水中的氢氧键形成，这一步会释放能量。只有当第二步的成键过程中释放出的能量超过一开始断键所需要的能量，燃烧才会以热的形式释放出能量。若这一能量的平衡颠倒过来，燃烧甲烷将可以制冷！

化学家利用热力学来记录这些单一过程中的能量变化，还利用它衡量反应中能量的总变化。为了达到这一目的，他们使用**焓**这一概念，以衡量反应前后的热量变化。这个名字源自希腊语，意为"内部的热量"。焓与能量这两个概念在技术上有明显区别，但是考虑到我们的目的，我们可以把焓当作储存于化合物中、可以热的形式释放的那部分能量。

在**放热反应**中，能量以热的形式释放，储存的焓减少。所有的燃烧过程都是放热的，在甲烷的燃烧中，甲烷＋氧气的焓减少至二氧化碳+水的焓，差值以热的方式逸散。化学家在评估燃料的效率时会考虑燃烧过程中伴随的焓值变化，焓值充足的燃料更受欢迎，因为从等量的燃料中可以获得更多的热。对焓与化学反应中释放的热的研究被称为**热化学**。它为我们了解食物与燃料做出了巨大贡献，也用于收集数据以进行热力学讨论。

除了燃烧，大部分反应也都是放热的，起始原料会转化成低焓值的产物，从而减少焓的总值。也许很容易理解许多反应都是向低焓方向进行的，正如那让我交叉手指的经验法则所表示的一样。然而，有这样一个难题，一个让19世纪的化学家极为困惑的难题：一些反应会自然地向增焓的方向进行。吸收热并增加焓值的反应被称为**吸热反应**。自然发生的吸热反应并不多，但是即便仅有一例也会让19世纪的所有化学家瞠目结舌，因为他们不明白：一个物体如何自然地上坡，在此例中就是增加焓值？

那时，他们并不知道熵的存在，他们按照字面意思理解经验法则，即事物自然地降低能量。化学家们现在知道了熵决定反应的方向，**只要熵增加**，反应既可以上坡增加焓值也可以下坡减少焓值。为了理解背后的原因，我们必须牢记熵是能量品质的指标。

当能量释放到反应烧瓶周围并逸散时，熵就会增加，所以应该很容易理解为什么放热反应很常见。然而，我们也需要考虑烧瓶内发生的变化。假设反应过程中，能量流入烧瓶：因为能量集中在局部、没那么分散、更容易获得，所以熵就会降低，能量品质随之变得更高。即便如此，设想与此同时在烧瓶内产生了极度的混乱。这时，尽管能量变得更集中，但是宇宙中的总熵还是有增加的可能性。如果出现这种情况，那么吸热反应也就会自然发生。

19世纪的化学家们失误的地方在于，他们认为反应会像牛顿的苹果自上而下坠落一样，沿着减少焓的方向发生；而21世纪的化学家们很清楚，反应会增加熵：混乱度增加，情况变得更糟。这两者通常会导向相同的结论，但在任何情况下，熵都是该考虑的性质。增加熵是变化的方向标，有时它会指向吸热的方向。如果你想继续从熟知的重力角度思考这一问题，与自然中发生的"下降"做类比，那么可以这样想：自然发生的变化总对应着能量品质的下降。

反应速率

现在，我们知道了反应会朝哪个方向进行：自然变化中的方向标会指向宇宙熵增加、能量品质降低的方向。两个问题随之而来。第一个是去向反应终点的速率，第二个是从哪一条路径到达终点。我会在这里解答第一个问题，在第四章解答第二个。

化学家们对化学反应的速率有极大的兴趣，因为只知道在理论上可以制得一种化学物质，而要花费千年才能获得一毫克是毫无意义的。对反应速率的研究被称作**化学动力学**。我们会看到，能量对于解释可观测到的各种反应速率极为重要。反应速率的范围着实宽广：一些反应可以在几分之一秒内完成（想想爆炸），有些却要经历数年（想想腐蚀）。

化学家们测量反应速率的方法很简单，即监测产物量随时间发生的变化。他们出于多种原因而进行这些测量。最基本的理由就是，想知道在任何给定时刻的浓度。还有一个更重要的理由，特别是对工业生产而言，就是他们想找到能以最优速率生成产物的条件。第三个理由是为了发现反应**机理**，即在原子层面上发生的从起始原料——"反应物"——转化成最终产物的变化次序。最后一个理由涉及的详细信息可以用这种方式获得：将一束分子射向另一束，然后监测撞击所产生的结果。

在此，我想谈论的是能量在决定反应速率中所扮演的角色。我们已经看到，反应的发生可能是自然的趋势，所以紧接着的问题就是，为什么所有反应不会瞬间结束。这个问题尤为重要，因为在很多情况下，缓慢可控的产物生成过程使得构建生命的精妙程序成为可能：如果生物反应都在一瞬间结束，那么我们会立即化为腐物。

化学家们已经确认了能量壁垒的存在，可以预防反应瞬间结束。通过测量温度对反应速率的影响，他们证实在反应物的原子重组为产物之前，分子至少需要获得一个最低的能量，被称为**活化能**。对气体中的反应来说，这一要求最容易理解，分子不间断地相互碰撞，不同能量产生的冲击不同。只有极快速运动的分子间的高能撞击才可以提供足够的能量，使得起初连接原子的化学键断裂并促进新的化学键生成。随着温度上升，分子运动变快，更多比例的碰撞会以等于或高于这一最低能量的条件发生，因此反应速率加快。某些反应中活化能垒极高，在常温下几乎不会有任何可以为反应提供足够能量的碰撞发生。氢气和氧气的反应就是一个例子：这两种气体在常温下可以永久地储存在一起，但是在高温下或当火花在局部提供足够的能量来启动反应的时候，就会发生爆炸。

这个最低能量的要求也适用于溶液中发生的反应，包括在那些由水组成的生命体内。在这种环境中，分子不会在空间内横冲直撞：它们会推挤着穿过液体、相遇，最后可能什么都没发生又推挤着离开。然而也会有这种可能性，当两个反应物分子在一起的时候，它们被周围的水分子剧烈地推动，其中的原子轻松分开并重新组合为产物。随着温度升高，发生足够剧烈的推挤的可能性也会增加，所以即使液态环境中的反应在受热时也会加快。例如，萤火虫在温暖的夜里会比在凉爽的夜里更快地闪烁；在厨房里，我们使用加热的方式诱发反应，我们称这种反应为"烹饪"食物。

在很多情况下，加入**催化剂**后反应会更快，催化剂是可以加快反应速率但本身却保持不变的物质。中文将催化剂译为"触媒"，它很好地诠释了催化剂所扮演的角色。催化剂通过为反应提供其他路径——另一系列的原子迁移与成键过程——一种活化能更低的路径来发挥作用。因为活化能更低，反应物间可以在常温下进行更有效的碰撞，所以反应发生得更快。催化剂是化学工业的命脉，其中，高效、快速地生产目标产物是至关重要的，整个工业的成功都依赖找到合适的催化剂。一个需要注意的要点在于，没有所谓的"万能催化剂"，每个反应都必须分别进行研究并设计出相应的催化剂。另一点在于，并不是每一个反应都可以被催化：在很多情况下，我们必须接受大自然对反应速率的设定。

我们身体的正常运转离不开催化剂。**酶**（enzyme，它由希腊语的zyme衍生而来，意思为使发酵）是具有催化剂功能的蛋白质分子，它们以极高的特异性和有效性控制着我们体内所有的化学反应。生命就是催化作用的体现。

平衡的本质

反应速率研究中一个很重要的课题就是，回答反应结束且变化不再明显的时候会发生什么。化学家将这种状态称为反应达到了**平衡**。在某些反应中，起始反应物分子完全消失，但在许多情况下，在起始反应物全部消耗完之前反应就会停止。对于后者，有一个具有重大经济意义的例子，即在"哈伯-博施法"中以氮气和氢气合成氨（NH_3）的反应，它是很多工业流程的源头，其中就包括世界上很多农用化肥的生产。在反应完全停止之时，只有一小部分氮气和氢气可以转化为氨，而且不论等待多久，或者投入多少催化剂，产量都不会提高。这一反应已经达到了平衡。

平衡只是反应在表面上停止。如果我们可以在原子层面上监测处于平衡态的反应混合物，就会发现其中仍然进行着化学反应。当一个反应处于平衡态时，产物仍然在持续生成，但这些产物会以同样的速率衰退回起始反应物。也就是说，化学平衡是一个**动态**平衡，其中，正向反应与逆向反应会以同样的速率进行，因此没有净变化。在合成氨的过程中，氨分子仍然在平衡态下不断地生成，但它们分解成氮气和氢气的速率与生成的速率相等，所以没有净变化。

化学平衡是动态的而不是静止不变的，这一事实的重要意义在于，平衡态仍会对条件的变化做出回应。所以，尽管我们体内的某些反应已经达到了平衡态，它们仍然会对温度及其他因素的变化做出回应，正是这种回应能力让我们得以存活。"内环境稳态"这一精巧而又复杂的平衡状态使得身体得以存活并保持警觉，它正是动态、灵敏的化学平衡的体现。对于工业上极为重要的氨合成而言，平衡是动态的而非静止的这一信息给化学家与工业带来了希望，也许可以操控化学平衡，进而提高氨的产量。这正是化学家弗里茨·哈伯（1868—1934）与化工工程师卡尔·博施（1874—1940）在20世纪初面临的机遇，他们适时地发现，只要选准催化剂并在高温高压下进行反应，就可以得心应手地操控反应平衡。就这样，他们为全世界供应了充足的食物。

章末结语与下一章

我们现在已经看到，能量既是可以促进化学反应的胡萝卜又是可以限制化学反应的小拖车，如此就可以澄清我在第二章末所提到的晦涩表述的含义。逸散于混乱中的能量等同于胡萝卜，它是化学反应的驱动力。反应物转化为产物的过程中需要跨越壁垒，能量又等同于小拖车，约束了向胡萝卜的自由飞跃。

至此，我还没有提到过任何关于反应物如何在实际中经过原子重排形成产物的过程。应用化学的核心就是弄清和理解这些变化并用它们带来令人惊叹的、几近魔法的转化，这也是我们这段旅程中的下一阶段。

第四章

反　应

　　每当有人想到化学的时候,他们会联想到与之相关的化学反应,那些会闪光、爆炸、改变颜色或产生刺鼻气味的反应。他们知道化学反应会在化工厂内进行,也知道燃料的燃烧和塑料、颜料或是药品的制造都是化学反应。也许某些人可以正确地想到烹饪会引发化学反应,大多数人可能至少会模糊地意识到我们本身就是复杂的试管,正是因为身体中进行的无数反应,我们才会活着。但是化学反应究竟是什么?当化学家们搅拌、煮沸液体混合物,把一种液体倒进另一种,以及在实验室里进行看似神秘的工作的时候,到底在发生什么?

　　实际上,他们是在诱导原子交换同伴。起始物质,"反应物",由处于某一种结合状态的原子构成;生成的物质,"产物",由处于另一种结合状态的相同原子构成。晃动、搅拌及煮沸使原子从一种结合状态转向另一种结合状态,将一种分子内的原子分开并促使它们形成其他类型的分子。在某些情况下,反应物中的原子会迅速按预想的新方式排列,然而在另一些情况下,化学家则必须策划、诱导，通过一系列精妙的步骤实现复杂的劝诱。燃烧和爆炸可能仅需要一个火花；但是生成一个复杂的网状药物分子却需要思考，运气，时间，以及细致、复杂、考究的规划。

化学实验室装满专业仪器，其中有很多用于确定某一反应的产物是否与化学家所期待或所认为的结果一致。我会在第五章中解释它们的一些作用。还有很多仪器直接用于诱导原子以及分离预期的产物和废料，类似于分离化学反应中的麦粒和麦糠。这些仪器包括试管、烧瓶、烧杯、蒸馏仪器、过滤仪器，以及各种加热器、摇床、搅拌器。尽管这些仪器五花八门（也很昂贵），但在化学家眼中，原子层面上只有几个反应：准确地说，只有四个。其实，我们值得停下细细品味，世界上所有不可思议的存在，包括那些源于自然的和人工合成的，都是由屈指可数的元素以及四种操作它们的方式造就的。

在本章剩余的内容中，我会向你介绍这四种最基本的反应类型。在某些情况下，它们串通一气，它们的合作乍一看去像一种新的反应类型，但当串通被拆穿时，就只剩下这四种基本反应类型了。

质子的转移：酸和碱

在物理学家确定"质子"之前，化学家就已经发现了这种基本粒子，但是化学家并没有意识到他们已经发现了它。第二章中曾讲过，质子是微小的、带单电荷的氢原子核。质子电荷量低（意味着它通常只被分子中与其相邻的原子松散地抓住），质量小（因而行动敏捷），这些特点意味着某个分子内的氢原子核，即质子，可能会突然脱离这个分子并嵌入其附近另一个更容易接纳质子的分子的电子云中。简言之，质子从一个分子转移到另一个分子是四个最基本的反应类型之一。

我们生活在酸与碱的世界里。尽管早期的化学家们熟悉不同种类的酸，但在很长时间之后他们才意识到酸是含有氢原子的化合物，其对氢原子核只有微弱的控制力，极易失去它们。就像酸的名字所暗示的一样（拉丁文acidus：酸的，刺激性的），酸因刺激性味道而为人们所认识。经历过这一危险测试并幸存的化学家们在那时（如今在我们品尝醋、苏打水和可乐的强烈味道时，它以更加美味可口的形式存在于我们的反应中）并不知道挑动他们味蕾的东西就是质子。对质子真正的认知在1923年才出现，英国化学家托马斯·劳里（1874—1936）和丹麦化学家约翰内斯·布仑斯惕（1879—1947）分别独立提出，酸是任何含有氢原子的分子或离子，它可以将其质子释放给其他分子或离子。并不是所有含氢的分子都可以释放出质子，因为质子可能深深嵌进电子云中，但也有很多类型的分子可以释放出质子，特别是如果分子中含有可以将质子周围的电子云吸走、使质子得以逃脱的其他原子。乙酸，即醋中的酸，就是这类化合物之一；其他例子还有盐酸（HCl）和硫酸（H_2SO_4）。如果你看到H写在化学式首位，这就表明它可以释放出质子并作为一种酸。（你可能会思考H_2O是不是酸，请稍等并继续看下去。）

一个巴掌拍不响。如果有一个质子的提供者（某种酸），那么就可以假设一定有一个质子的接受者，一个可以让释放出来的质子附着并钻入其电子云的分子或是离子。碱液就是这样一类物质（名字源于阿拉伯文al qaliy，译为灰，因为木灰就是碱液的一种来源）。

在过去，对碱液的测试与对酸的测试一样危险：在测试中，碱液会有肥皂的感觉。我们现在知道碱液可以将脂肪变成肥皂，所以在测试中，测试者手指上的脂肪会转变为肥皂。当然，如今化学家们有了更安全且先进的测试方法。碱液将脂肪变成肥皂的根本原因在于，其中含有**氢氧根离子**（OH⁻），这类离子可以吸引和结合质子，并在这一过程中形成水分子（H_2O）。

我必须在此介绍一个专业上的细节。如今，化学家们把可以接受质子的分子和离子称为"碱"。所以，OH⁻是碱。他们保留"碱液"这一概念，用于描述溶于水的碱。例如，氢氧化钠（NaOH）溶于水，形成 Na⁺（钠离子）和 OH⁻（氢氧根离子），所以氢氧化钠是碱（OH⁻），其溶液是一种碱液。在这之后，我会使用"碱"这一术语，因为它比碱液更宽泛（某一分子或离子不需要在水中才可以被称为碱）。

为什么要使用"碱"这个名字？当盐酸与氢氧化钠溶液反应的时候，酸提供的质子可以转移而与氢氧化钠提供的氢氧根离子（OH⁻）结合，生成食盐（氯化钠）和水。在另一种情况下，当硫酸与氢氧化钠溶液反应的时候，酸提供的质子转移而与氢氧化钠提供的氢氧根离子（OH⁻）结合，生成硫酸钠和水。我们以相同的碱——氢氧化钠——为基础，结合成不同的化合物，氯化钠和硫酸钠：碱由此得名[①]。

顺带一提，氯化钠和硫酸钠都被称作**盐**，这一类离子化合物都是由酸与碱反应生成的，得名于一个常见的范例，也就是常见

[①] 碱的英文为 base，有基础之义。

的食盐,氯化钠。这种命名方式是化学中一个普遍的特点,某一例化合物的名字会启发一整类相关化合物的命名。

大部分化学反应都是酸与碱之间的反应,这类反应的共同特点为,质子从酸转移到碱。其中最重要的是那些发生在有机体内的反应,包括玉米、橡树、苍蝇、青蛙,还有我们自己,因为很多由酶控制的生化反应都属于酸碱反应,例如涉及食物和呼吸的新陈代谢。其实,你可以把生命看作一个漫长的、高度复杂的滴定过程!

质子转移、酸碱反应的重要性在于,增加一个带正电荷的质子会扭曲碱的电子云,由于分子中与其相邻原子的电子云被拉向质子,使得暴露出的原子可能被其他原子进攻。所以,质子转移可以使原子以及维持原子间联系的化学键做好进攻的准备并进一步反应。我们体内酸碱反应的一个主要作用就是为这类进攻反应做好准备,而酶会为消化或修饰小分子做好准备。

在之前的内容中,我猜你想知道水的化学式 H_2O 首位的 H 原子是否暗示它是一种酸。答案是肯定的。你喝水的时候其实就是在喝纯度接近100%的酸。水同时又是一种碱。你也应该知道,喝水的时候也意味着在喝纯度接近100%的碱。我需要解释这一令人大跌眼镜的真相。尽管令人震惊,但承认水既是酸又是碱这一事实,对于化学家们思考水、水所形成的溶液以及水所进行的反应的方式都极为重要。

请你将自己想象为一杯水中的某个水分子,周围是密密麻麻的其他水分子。你的一个氢原子的质子可以脱离你并附着在一个邻近的水分子上。这一转移意味着,作为一个质子提供者,你是一种酸。接受了你的质子的邻居表现为一种碱。你失去一个质子,成了一个氢氧根离子（OH⁻）；你的邻居获得了一个质子,成了一个**水合氢离子**（H_3O^+）。这个外来的质子就像一个烫手的土豆,立刻被传递给你邻居的邻居们。类似地,你携带的负电荷可以让你从一个邻居那里获得一个质子,使你再次变回H_2O。这一永不停息的、混乱的质子传递过程,由OH^-变为H_2O再变为H_3O^+的快速转变,会在整个液体中持续进行。OH^-和H_3O^+的实际浓度非常非常小,所以当你看到一杯水的时候,你应该这样想,里面绝大部分是H_2O,其中仅有几个OH^-和H_3O^+。随着质子快速四处转移,这些分子和离子的身份会不断地改变。每个H_2O都是一种酸（质子提供者）,同时每个H_2O又都是一种碱；这就是我说水既接近纯酸又接近纯碱的原因。

电子转移：氧化和还原

物理学家约瑟夫·约翰·汤姆逊于1897年发现了电子。在此之前,化学家们已经不经意地反复转移电子长达几十年之久,使用电子转移的最具代表性人物为迈克尔·法拉第（1791—1867）；显然,即便是他,也不知道自己到底在做什么。四种基本反应中的第二种就是电子从一个分子转移到另一个,很多化学反应都源自这一小小基本粒子的转移。比如,电子转移是炼钢等大工业的基础。电子转移也会通过腐蚀作用损坏钢铁制品。

我需要首先介绍"氧化"和"还原"的概念,然后解释电子转移在其中扮演的角色。氧化与其字面意思相同,即与氧气反应。然而,尽管科学中的某个概念可能起源于日常用法,但它通常会以概化的方式涵括进同一图景的其他部分。刚才我们了解到,"盐"这一概念源自一个单独的范例,之后被推广用于一整类相关化合物。氧化这个概念也是如此。

举一个简单的例子。我们大部分人都见过镁（Mg）条在空气中燃烧发出耀眼的光。在这一反应中，金属镁与氧气结合生成氧化镁，它是一个由 Mg^{2+} 和 O^{2-} 组成的离子固体。反应中释放的能量以光和热的形式散发出来。然而，值得注意的要点是，金属中的每个 Mg 失去两个电子，变成带两个电荷的 Mg^{2+}。当镁在氯气中燃烧并生成氯化镁时，会发生类似的反应，但这一反应并不常见。这一化合物由 Mg^{2+} 与 Cl^- 组成。与第一个反应类似，反应中发生的重要变化是每个 Mg 失去两个电子变成 Mg^{2+}。氧气没有参与第二个反应，但两个反应中都失去了电子。如今，化学家们把第二个反应简单地归为广义的氧化反应，并将氧化定义为**失去电子**。有时很难确定是否失去电子，比如在烃类燃料的燃烧过程中，但化学家们有许多辨别的方法，每当失电子反应发生的时候，尽管氧气本身可能不会以任何方式参与其中，他们也会称之为氧化反应。

我们在前面讨论酸碱反应的时候，曾提到过一个巴掌拍不响：如果有提供质子的一方（酸），就一定会有接受质子的另一方（碱）。同样的道理也适用于电子转移反应，氧化过程失去的电子最后一定会出现在某处。这就对应了还原反应。

在过去（我在有意地模糊时间），还原特指将金属从其相应的矿石中提取出来：金属矿石被**还原**为金属。比如，这一过程曾发生在工业革命的一个巨大笨重的象征符号，即高炉内，其中铁矿石（铁的一种氧化物）与焦炭、一氧化碳反应生成熔融的铁（Fe，源于拉丁文中的 ferrum 一词），熔融的铁从高炉底部滴流而下，后续成为不同类型的钢。氧化铁由 Fe^{3+} 和 O^{2-} 组成。金属铁由Fe组成。牢记这些后，就很容易明白还原矿石的过程中发生了什么：电子与Fe^{3+}结合，由此中和了它的电荷，生成Fe。

如今，还原被定义为电子与原子结合的反应，但它可能与将矿石还原成金属的过程完全无关（除了上述特点之外）。所以，镁在氧气中燃烧，氧分子获得镁在氧化过程中释放出的电子，从而变成O^{2-}：氧被还原。在镁被氯气氧化的过程中，氯分子获得释放出的电子，变为Cl^-：氯被还原。每当释放出的电子转移到某个原子上的时候，那个原子就被称为被还原了。

现在，电子转移反应中的两个一半反应合在了一起：氧化反应（失去电子）总是伴随着还原反应（得到电子）。化学家意识到了这两个一半反应合在一起的必要性，不会单独提起氧化反应或是还原反应，而是称"氧化还原"反应。（迄今为止，他们还没将这种压缩式的命名方法拓展而提出"酸碱"反应，将酸与碱两个一半反应合并在一起。）

氧化还原反应极为重要。我们已经看到，将铁矿石中的铁提取出来的反应位于炼钢链的最前端。腐蚀与提取铁的过程相反，钢铁制品会在一种我们称为腐蚀的氧化还原反应中损失：铁被水和空气中的氧气氧化，转变回铁的氧化物。驱动我们的机动车的燃烧反应也是氧化还原反应，其中，烃类燃料在氧气作用下被氧化为二氧化碳和水（氧气自身被还原）。

氧化还原反应为我们的笔记本电脑、平板电脑、手机，以及逐渐增多的机动车内的电池提供电力。电池作为移动电源对现代社会极为重要，所以有必要用一些时间理解它们运行的基本原理以及它们怎样通过氧化还原反应提供动力。

我们已经看到，电子在氧化过程中被释放，在还原过程中被获得。电池内,电子的释放与获得在空间上是分开的。电子首先被释放到电池中某一区域内的一个电极,即一个金属接触点,再经过一条电池外线路,然后在电池中另一区域内的第二个电极附近,与发生还原反应的物质结合。这样,协同的氧化反应与还原反应,即氧化还原反应,可以持续进行。以这种方式,电子从一个电极流向另一个电极,为连接到这一电池上的任何电器设备供应能量。现代电池通过各种氧化还原反应引发电子流动,这些电池大至机动车内笨重的铅蓄电池,小至笔记本电脑、平板电脑以及手机内轻便的锂离子电池,各种各样。

将电子经过电极通入反应混合物中,可以强制氧化还原反应逆向发生。这个过程被称为**电解**,即以通入电流的方式引发化学反应。电解是从氧化铝中提取出铝(Al)的主要方式。一股强大的电流被通入电解池内,其中含有溶解在特殊溶剂中的氧化铝,进入电解池的电子被强制与Al^{3+}结合,形成Al。电解也用于提纯铜以及将金属,例如铬,镀到其他金属表面。

与酸碱反应中的质子转移不同,氧化还原反应中电子转移的一个特别之处在于,由于电子与成键密切相关,电子从一个分子转移到另一个分子时会拖动几个其他的原子。我们之前已经见过了几个例子,但并没刻意提起,在燃烧反应中,随着烃分子的氧化,电子在分子间转移的同时,碳、氧及氢原子也被拖来拖去,最终烃和氧分子重组为二氧化碳和水分子。这一不同之处在有机反应中尤为重要,巧妙使用氧化还原反应中原子的拖动可以构建出各种复杂的分子结构。

在某种程度上,正是因为电子在转移时可以拖动原子这一行李,氧化还原反应在生物界也极为重要:它们让生物圈保持

(也包括你这一小部分)生机勃勃。光合作用,即绿色植物吸收阳光、合成碳水化合物的过程,就是一连串的电子转移反应,考虑到原子拖动,其最终的整体效果就是水中的氢原子及二氧化碳中的碳原子、氧原子被用于构造碳水化合物,包括淀粉和纤维素。在被我们称为消化的有机腐蚀过程中,曾以氧化还原方式生成的碳水化合物,经过被我们称为呼吸和新陈代谢的一系列氧化还原反应,其中的碳原子与氢原子被开采出来。

自由基反应

第三种反应类型在自由基碰撞的时候发生。你需要了解,**自由基**是带奇数个电子的分子。我们已经看到,化学键形成时,电子会配对在一起,所以自由基中除了一个电子外,其余所有电子都已配对并将原子维系在一起,未成对的电子只剩下一个。自由基通常写作R·或者·R,其中圆点代表未成对的单电子。

大部分自由基都有强烈的反应性,在自然环境下不会存在过久。在某些情况下,两个自由基可能会发生碰撞,由于它们未成对的电子可以结合成对,所以两个自由基可以聚集在一起,形成一个含有偶数个电子的普通分子:R·+·R→R—R。这种反应可以发生在火焰中,因为高温可以拆开分子,使成键电子对分离,所以火焰中富含自由基。事实上,有一种阻燃剂就是在加热时产生自由基。这些自由基,以·X表示,可以结合在传播火焰的自由基上,消除它们的化学活性,R·+·X→R—X,使火焰熄灭。

有些自由基具有重要的商业价值,因为它们被用于塑料生产。这一过程被称为**聚合**,其基本原理为:当一个自由基R·进攻一个普通分子M的时候,它们可能会结合到一起。然而,结合产生的分子RM·带有奇数个电子，所以同样是一个自由基。这一自由基可以继续进攻另一个分子M，与之结合。结果仍是一个自由基，表示为RMM·。换句话说，这属于一种**链式反应**，一连串的过程可以无限地进行下去，生成一个长长的RMM⋯M·自由基，直到两个这样的自由基碰撞，通过电子对粘在一起，如此终止反应链。

那些无处不在的塑料，像是聚乙烯、聚苯乙烯及聚氯乙烯，就是以这种方法制成的。在聚乙烯的例子中，分子M（此处被称为**单体**）是乙烯（$H_2C=CH_2$）。经聚合反应可以得到一条由数百个—CH_2CH_2—单元组成的长链，即**聚合物**。化学家们发现以乙烯的不同衍生物——例如$H_2C=CHX$，其中X可以是一个原子团——为起始原料，经过聚合反应，可以得到具有不同性质的聚合物。所以，若X是一个苯环，聚合产物就是聚苯乙烯，若X是一个氯原子，聚合产物就是聚氯乙烯。为了得到不粘的特氟龙®材料，乙烯中的所有氢原子都被替换成氟原子，因而它正式的名称是聚四氟乙烯（PTFE）。

另一种类型的酸：路易斯酸

第四种也是最后一种基本反应类型可能第一眼看上去有些晦涩难懂，但它是一类极其重要的反应。前一小节中，我们了解到，如果两个自由基每个都带一个电子，它们的电子就会配对，就可以形成化学键。在最后这种类型的反应中，一个分子可以提供成键所需的**两个**电子，另一个分子可以接纳这两个电子。我们可以将这种反应类型表示为A + :B → A—B，B旁边的两个圆点代表即将与A共用的一对电子。这种反应类型以美国化学家吉尔伯特·牛顿·路易斯命名,被称为**路易斯酸碱反应**,他是首位发现这两种分子的化学家,后来因它们而丧生。[他在摄入氰离子(CN^-)后死亡,CN^-是一种以此种反应起作用的毒药。]它们之所以被称为"酸碱"反应,是因为它们与我之前提到过的酸碱反应高度相似,其中质子从酸转移到碱。实际上,它们可以被当作对普通酸碱概念的泛化,但我不打算带你们去领略这一思路通往的那幅精彩景象。

路易斯酸碱反应的一个作用在于为世界带来了色彩。由这句话开始,我将为你们介绍**过渡金属的络合物**,它们通常颜色鲜艳,是以我所说的路易斯方式形成的。你血液中的血红蛋白就是一个例子。

过渡金属是位于元素周期表中间狭长区域内的元素,包括铁(Fe)、铬(Cr,这个名字预示颜色,因为chroma在希腊语中表示颜色)、钴(Co)、镍(Ni)等。这些元素形成的离子,例如Fe^{2+}和Co^{3+},通常被六个独立的小分子或离子围绕并与之成键,例如H_2O、NH_3和CN^-。这些分子和离子被称作**配体**,它们与过渡金属离子形成的完整聚集体被称为**络合物**。每个配体提供一对电子,与中心离子成键而形成络合物,所以金属离子表现为路易斯酸(A),每个配体表现为路易斯碱(:B)。

在水中,过渡金属离子通常被作为路易斯碱的六个水分子包围。若溶液中加入另一种路易斯碱,它可能会挤出一个或更多个水分子,并占据它们的位置。由此产生的络合物的电子结构可能与原络合物完全不同,而呈现出鲜艳的颜色。很多颜料与染料都是以这种方式形成的络合物。

呼吸是一个路易斯酸碱反应。在血液中,氧的载体是血红蛋白,它是一种嵌有四个铁离子的巨型蛋白质分子。每个铁离子都被蛋白质框架上的四个氮原子牢牢固定,这四个氮原子位于正方形的四个角。铁原子与氮原子之间形成的键是路易斯酸碱作用的结果,Fe^{2+}充当路易斯酸,每个:N充当路易斯碱。当你吸入空气的时候,这一由路易斯酸碱反应生成的实体会参与另一个路易斯酸碱反应,其中,一个氧分子会表现为路易斯碱,它的一对电子可以与血红蛋白分子中的一个Fe^{2+}形成化学键。一旦捕捉到宝贵的氧,它就会被血液输送到身体内部,参与其他反应。

　　由一氧化碳中毒引起的窒息是另一种路易斯酸碱反应。此例中,一氧化碳分子(CO)可以篡取氧气在络合物中的位置并通过路易斯酸碱反应与血红蛋白中的Fe^{2+}结合。一氧化碳的结合能力比氧更强,所以篡位后的一氧化碳可以阻止氧的结合,如此导致氧不能被输送到需要的地方,受害者因而窒息。这一窒息过程发生在分子层面,与呼吸道受阻大不相同。我之前提到过的氰离子(CN^-)的中毒机制与此类似,不同的是,它阻碍的是呼吸作用后期的一连串电子转移反应。

有机化学的精妙之处

　　有机化学家们使用这些基本反应类型的时候,就像魔术师,或指挥作战的将军。他们必须如此,因为他们希望构建的分子通常是由原子组成的精妙结构,一个原子错位就可以使药剂失去活性或使研究倒退数月。在过去几十年有机化学的发展过程中,化学家们已经累积了丰富的经验,他们知道如何诱导原子合理地排布以满足他们的需求,有时这一过程需要经过一系列反应,包括数十步,其中任何一步失败了都可能将来之不易的化合物变成化学家眼中一块黝黑的、毫无用处的焦油。这些步骤通常以发明它们的化学家的名字命名。在设计制备方法的时候,电脑软件也会发挥作用,就像它用于制定工程项目的流程一样。

工程项目的比喻可以更进一步。在项目中未完成部分的附近施工的时候,其中未完成的部分可能需要保护;类似地,一个未完全构建好的分子可能会有薄弱的地方,不进行保护就会成为反应中心,导致生成不需要的产物。所以,化学家们有时会将一个小原子团连接到分子内的某个区域,或是用于防止邻近区域受到进攻,或是用于隐藏与其相连的原子。起保护作用的原子团在后续步骤中可以除去,类似于拆除建筑物的防护罩。

我仅举两个例子,说明有机化学家如何构建分子,这种分子将作为药物、染料或人造香精接受测试。这两个例子都属于**取代反应**,在这类反应中,一个原子或原子团会取代分子中已有的一个原子或原子团。在每种情况下,引入的反应物都可以检测到电子云较薄或较厚的区域,从而确定目标原子。如果电子云较薄,带正电荷的原子核就会透出亮光,带负电荷的反应物分子就可以像制导导弹一样飞向它。这类检测原子核电荷的反应被称为**亲核取代反应**。相反,如果电子云较厚,那么带负电荷的电子就会力压带正电荷的原子核,带正电荷的导弹就会飞向该区域。这类寻找电子富集区域的反应被称作**亲电取代反应**。

当计划构建一个分子的时候,化学家需要思考电子云在分子中分布的方式,据此选择反应物分子。他们需要巧妙地设计反应路径,因为新加入的原子团可能吸引电子,使它们远离某一区域,也可能将电子推向某一区域。通过这种调整电子云的方法，化学家有理由相信反应物分子可以准确地飞向目标原子，形成新的化学键。

至此为止，我希望你开始感受到化学的精妙之处，化学家以此创造出各种此前宇宙中并不存在的物质。虽然很难从这段简短的介绍中理解化学家设计反应时的具体细节，但我仍期待你可以感知到他们工作背后的谨慎周密。

第五章

技 术

走进任何一间现代化学实验室,你都会发现里面有各种实验仪器。炼金术士会认出其中的一些仪器,对剩下的则完全陌生。盛装液体的容器只有几种形状,且大部分有明确的历史渊源。现代**分析**在语义上指对物质的分解,在现代操作中指对组成物质的识别以及对组成物质的量与浓度的测定,它使用精密复杂的电子仪器,而且通常是自动化的。分析并不是实验室里的唯一工作,与之相对的**合成**在语义上指将物质组合在一起,但在实际操作中,它指以简单的成分创造出想要的物质形式,因此它同样也是化学工作的一个重要组成部分。

经典实验仪器

我不会再赘述烧杯、烧瓶、试管这类仪器,因为它们盛装、混合液体的用途是显而易见的。然而,还有一些容器用来转移定量液体,或是为了像在厨房里一样按照精确的配方进行操作,或是作为定量测量的一部分。后一用法的一个例子为,使用移液管（英式英语为pipette，美式英语为pipet；意为小管）和滴定管（英式英语为burette，美式英语为buret；buret源于法语中的小花瓶或水罐，不过实际中的不同于想象中的）进行一项化学中的经典实验操作，即用一种碱液滴定（titrate）一种酸液，从而确定酸或碱液的浓度。（为什么使用titrate？因为titre在法语中是测定或是测试的意思。）移液管用于将定量的碱液转移到锥形瓶中；滴定管用于将酸液逐滴加入锥形瓶中，直到颜色发生变化或检测器的电信号指示碱液完全被中和。记录下来从带刻度的滴定管内流出的酸液的体积并了解其浓度，就可以确定碱液的浓度。

另一类仪器与分离物质有关，可用于提纯或分离产物。若产物是两种溶液混合生成的固体沉淀物，一个最直接的分离方法就是"过滤"：将反应后的溶液通过一个细筛。另一个经常用于分离液体的方法是"蒸馏"：将液体混合物煮沸并冷凝蒸气；混合物中挥发性更强的组分会先沸腾，冷凝得到的液体可被收集或丢弃。

此外，还有一种极其精密的分离方法，就是"色谱法"。这项技术的诞生与命名源于一个简单的观察，某滴溶液，例如取自某一开花植物的溶液，可沿吸水纸扩散并形成不同颜色的条带，这些条带可被辨别或收集。尽管这一名字沿用下来，但这项技术已得到极大改进。如今，在一套典型的流程中，待分析的样品要通过几米长的细管，细管内涂有一层用于吸附的固体。混合物中的组分会不同程度地黏附（专业名词为"吸附"）在内层表面，尽管它们都可以通过细管，但会在不同时刻出现，所以它们可以被分别收集起来，并用其他方法测定。这个方法已被用于分离使水果呈现风味的化合物，除此之外，还可以使用一种更专业的色谱法,例如在安检处,搜找出易爆品。

光谱学

在实验室内,让炼金术士觉得更有趣却又不解的应该是电子仪器,它们只有屏幕、旋钮,而不直接标示出用途。许多电子仪器的原理都是**光谱学**(spectroscopy)。这一术语源于拉丁文中的 spectrum 一词,意即表象,或观看某个表象的过程;但现在"观看"比目测更复杂,"表象"也不同于其日常含义。

我要从原子光谱学说起。当某个元素被气化与加热的时候,原子中的一个或更多电子可能会从其正常分布中发射出去,短暂悬浮于原子上方,随后回落至正常的电子云中。电子回落可以产生一个脉冲,射向我们设想中环绕原子的真空,脉冲随即产生一束光脉冲,即**光子**。光子的颜色取决于电子回落时释放出的能量,高能量会产生紫外辐射波段的光脉冲,低能量会产生可见光波段的光脉冲。原子中的电子可以以元素所特有的能量态存在,当电子从激发后的能量态回落时,它们就可以产生相应颜色的光子。我们都很熟悉黄色的街灯,这一颜色的光是由于受激发的钠原子回落到正常状态时释放光子而产生的;类似地,霓虹灯招牌呈红色则是由于氖原子的电子回落到正常状态会产生红光。通过观察颜色模式并"记录光谱",我们可以确认某种元素的存在。

分子内的电子表现出类似的行为特征,但需要使用一种稍微不同的方式来监测它们可能处于的能量态。我之前描述的原子光谱学利用的是光的发射,分子光谱学与之相反:它利用的是光的吸收。

通过样品的光线可以被想象为一束光子流。如果其中的一个光子与分子碰撞，以相配的能量将分子激发到某一更高的能量态，光子就会被吸收。这部分光子从进入的光子流中移除，即被"吸收"，从而降低了光束强度，这一变化可用某种检测器记录下来。为了记录完整的吸收光谱，需要系统地改变入射光的颜色，同时监测通过样品后的出射光的强度。因为每个分子都有特征性能级，所以它们的吸收光谱都是独一无二的，可当作确定分子的良好指标。

上文提到，分子内常态下分布的电子受到激发可以吸收光子。这需要消耗很多能量，尽管很多分子可以吸收可见光（正因如此，我们的世界充满色彩），但是我所描述的光谱通常利用紫外辐射。故而，这项技术被称作"紫外-可见分光光度法"。一个极其相近的方法使用红外辐射波段的光子，它比可见光和紫外光光子的能量更低。这些光子可以激发分子的振动，而不是电子的分布。所以，"红外光谱"说明了分子振动可以被激发。这对分析复杂分子内的原子团有重要的作用，例如，CH_3 可以吸收某一能量而摆动，而 CO 可以吸收另一能量而摆动。

核磁共振法

也许光谱分析法中最重要的一种就是"核磁共振"（NMR）。"核"字出现时通常预示着危险，这是它没写进"磁共振成像"（MRI）这种医学检查方法的原因，其实 MRI 是由 NMR 衍生出的一种技术。相较于大众，化学家总体上不易对"核"这个字感到恐慌，于是保留了它的用法，因为他们清楚在这种语境中它与放射造成的危害完全无关。

核磁共振中的"核"指的是任何一种原子中的任何一种原子核，但是我将专门讲述其中最常见的原子核，质子，即氢原子的原子核。质子像地球一样围绕自身的旋转轴自旋，这个旋转的电荷相当于一个小磁棒。它可以顺时针或逆时针自旋，依据自旋方向，相应的小磁棒北极会朝上或朝下。当自旋质子处于磁场（实际上是电流通过超导线圈产生的强磁场）中时，两种自旋方向对应了不同的能量，一个频率适当的入射光子可将一个取向朝上的质子（低能量）翻转为取向朝下的质子（高能量）。光子的频率符合两种能量的差值，这就是"共振"；当我们将收音机调为远处无线电发射器的频率的时候，我们其实是在做同样的事情。自旋取向翻转后，入射的光子流衰减，从而检测到降低的光束强度。两种状态下能量相差不大，这项技术使用射频辐射的光子，这一波段位于调频无线电信号的高频端（100兆赫左右）。

翻转质子的取向可能看似毫无意义。这项技术不可低估的优势在于，共振发生的准确频率取决于质子，确切地说，取决于氢原子的原子核在分子内所处的位置。与碳原子相邻的氢原子核相较于与氧或氮原子相邻的氢原子核，它们在不同频率发生共振，所以共振吸收谱，即"核磁共振谱"反映了分子内所有氢原子邻域。

不仅如此。在同一分子内位于氢原子中心的那些小磁棒可以相互作用，并改变彼此的能量。这种改变可以进而影响质子的共振频率，并出现特征吸收模式，对识别分子起到重大作用。

碳原子核不自旋，所以它不具有磁棒的特性，在核磁共振谱中不可见。这是幸运的，否则，即便是一个很简单的有机分子，它也会表现出极其复杂的核磁共振谱。然而，将普通的碳12替换成它的同位素碳13，碳原子就可以被观测到，因为碳13

的原子核内含有一个额外的中子，带有磁性。所以，理智地将碳12替换为碳13也可用于绘制碳原子在分子内的位置图，如此就可以确定分子的身份与结构了。

质谱法

除此之外，还有一种完全不同的光谱仪，它不利用吸收或发射，而是从全新的角度鉴定分子。"质谱仪"可以把分子轰击开，并对产生的分子碎片称重，然后根据分子碎片的质量推测出分子的成分。

利用电子轰击可以引发分子碎裂，电子撞击到分子后，使维系分子的电子云发生扭曲，产生很多带电荷的分子碎片。这些带电碎片经过电场加速，从强磁场的两极之间穿过，轨迹会发生偏转，偏转的程度取决于质量大小与磁场强度。特定质量的碎片会落在探测器上并发出信号。随着磁场发生变化，不同质量的碎片会由探测器检测到，据此绘制成的谱图是由不同质量的碎片组成的"质谱"，对质谱进行分析可以得出母分子的结构，就像拼凑花瓶碎片可以复原打碎的花瓶。

X射线衍射法

生物学中，结构对功能极为重要。结构也几乎是化学的全部，在化学与生物学的融合之处更是如此，化学家们为研究和阐释酶这种大蛋白质分子的作用机理做出了贡献。尽管酶对调节构成生命各个方面的化学反应极为重要，但是它们并不是有机体（包括人）内唯一的重要成分。DNA使遗传成为可能，刚性蛋白和骨骼提供了支架，探测和传递信息的分子使认知和思考得以实现。整个有机体的运转都由渗透其中的分子来调节。

用于发现结构的最重要工具之一就是"X射线衍射法",因为这一方法总是以目标物质的晶体为研究对象,所以又被称为"X射线晶体学"。这项技术成了诺贝尔奖的源泉,从威廉·伦琴发现X射线开始(于1901年获得第一届物理学奖),然后是威廉·布拉格与他的儿子劳伦斯·布拉格(1915年),接着是彼得·德拜(1936年),后来是多萝西·霍奇金(1964年),1962年莫里斯·威尔金斯(却不是罗莎琳德·富兰克林)获奖将其推向顶峰,这为詹姆斯·沃森与弗朗西斯·克里克对DNA双螺旋结构的构想奠定了基础,此项成果对理解遗传、治疗疾病、抓捕罪犯都有深远意义(1962年他们与威尔金斯共同获奖)。如果有一项可以让生物学与化学融为一体的技术,那么它一定是X射线。这份名单的另一个令人惊叹的特点就是,以下三个科学领域都授予了奖项:化学、物理学,以及生理学和医学,这就是这项技术应用的范围和它为我们带来的光明。

为了理解这项技术的基础,我们必须知道,X射线是波长极短的电磁辐射,它与光类似,但是波长只有光的千分之一(大约100皮米,差不多等于原子的直径)。第二个重要信息是,与所有的波一样,X射线可以相互干涉:波峰相遇的地方会更亮,波峰遇到波谷的地方会变暗。当某个物体被放在X射线的路径上,物体会使X射线发生散射,由一个分子的不同部分散射的X射线会以不同路径到达探测器,所以会以不同的方式相互干涉。这项技术名称中的"衍射"就源自处于X射线路径上的物体引发的干涉现象。

在X射线衍射法的测定过程中,一小块晶体样品在X射线的路径上旋转,探测器环绕样品移动,同时检测相长干涉产生的闪光。采用数学方法就可以用采集到的海量数据确定样品中原子的排列方式。如今,这项技术已广泛地实现自动化,集成电脑可以控制数据采集与解析。

测定中最具挑战性的部分就是制备对该技术十分关键的晶体,尤其是制备作为最主要研究对象之一的大分子。然而如果只需要鉴定一种化学物质,例如一种矿石,那么就可以使用一种相对简便的技术。将待测样品的粉末铺在实验板上,当一束X射线照射到粉末上时,可以采集到它的特征性"粉末衍射图样",然后通过比对图样库来确认这种物质。

为什么使用X射线?当辐射波长与可发生衍射的结构尺度吻合时,就可以得到衍射图样,在我们的例子中,原子组成的结构产生衍射。因为X射线的波长与分子内原子间的差值恰好吻合,所以使用X射线是最佳方法。

绘制表面

尽管固体内部令人着迷,但反应通常发生于固体表面。比如,在催化过程中,反应通常在一个不参与其中的物质的作用下以一种机制加速发生,在这一机制中,反应物结合于固体表面,有时会发生断裂,从而准备好与其他反应物进行反应。化学工业的存在归功于催化剂,因此研究固体表面发生的反应尤为重要。

尽管表面是固体与外界相连的窗口,但很难对它进行研究,直到几年前突然出现了一项令人激动的新技术。这项技术极为灵敏,可以描绘出固体表面的单个原子以及贴附于表面的分子。它有两种方式:"扫描隧道电子显微镜技术"(STM)和"原子力显微镜技术"(AFM)。

第一眼看去,STM并不可行。探针被拉伸出极细的针尖,连续横向扫描一整块待研究固体的表面。针尖与表面之间的电流流动可以被监测到并绘制于显示屏上:针尖经过表面凸出的原子,导致电流激增,在屏幕上显示成一个峰。"隧穿"量子力学效应(这项技术名称中的"隧道"由此而来)让这项技术成为可能,电子可以跨过禁区,在此指固体表面与针尖之间的缝隙。隧穿效应对此缝隙宽度极为敏感,所以横向扫描表面可以检测到表面自身在原子尺度上的变化,也可以极为详细地显示、描绘出贴附在固体表面上的分子的形状。

有一种宽泛的说法,原子太小所以看不见,然而只要我们将"看到"的视野扩大至隧穿电流变化的图像,那么STM就会否认这一说法,并为我们提供单个原子和分子的令人叹为观止的图像。即使洁净的固体表面也会令人惊叹,原子堆积的地方类似火星表面的山峰和悬崖,原子缺失的地方好像深渊。今日,固体表面终于可以被仔细、直接地观察研究。

原子力显微镜技术直接作用于固体表面。与之前被动观测相反,探针可以在表面四处移动,将原子从某处推向另一处。除了可被用于在固体表面移动C_{60}"巴克球"来玩"纳米足球"的游戏,精细地操控探针还可以排列单个原子。若探针涂有某种分子,则移动探针就可以在固体表面刻画出各种图样,并在纳米尺度上构建出各种结构(参见第七章)。

计算化学

近几十年来,计算机彻底地改变了化学,就像它从各个方面彻底改变了生活。除了最原始的实验操作外，几乎全部操作都由计算机控制。就像我们看到的，计算机是X射线晶体学的基本组成部分，解释衍射图样也需要计算机。它们也是现代核磁共振技术不可或缺的，核磁共振使用特殊技术来观察图谱，这需要大量数学运算来得到实际图谱。除此之外，计算机也有发挥自身特点的应用，即计算分子结构并呈现其图像。这就是"计算化学"领域。

尽管计算机硬件有如此高程度的发展，许多分析可以在平板电脑甚至手机上进行，但是与气象学家和密码学家一样，化学家同样属于对性能要求最高的计算机用户。

计算化学内的一个分支将研究主题追溯到对分子内电子云分布的量子力学描述，并致力于计算其分布。这类计算中包含了很多数字运算和各种各样的近似算法。尽管最终结果只是一串关于分子内电子云密度的数字，但通过以图像呈现电子云的方式，这些数字被赋予生命、可被理解，让化学家们可以评估分子可能的行为。可以绘制出密集与稀疏的电子云区域极大地促进了一个非常重要的应用，即评定分子的药物活性，以及在进行动物实验前对可能有药物活性的化合物进行初步筛选。

与计算化学关联紧密的第二个分支研究的是，蛋白质如何折叠从而形成具有活性的形态。蛋白质分子是以化学键连接分子（氨基酸）的方式形成的一条长链，它可以折叠形成螺旋和片层结构，进而折叠形成具有足够刚性的结构从而满足其功能。尽管我们很好地掌握了同一分子内不同部分的相互作用力，但是理解这些不同的作用力如何共同将长链扭转形成最终形态仍然是一个悬而未决的难题。自然可以实现这一过程，但我们仍不理解它如何发生。作为解决这个难题的一项手段，计算机被用于追踪分子链内的连接如何翻转扭动而形成最终形态，从而理解自然怎样不假思索地实现这个过程。

计算机也被用于研究小分子的行为。一个分子被抛向另一个（在软件的想象中），人们通过计算观测化学反应中分子紧密结合，即分子撞击、旧键断裂、新键形成时发生了什么。

现代合成方法

现在我想将关注点从研究转向合成，转向一种当下流行的独特的合成方法，用这种方法，化学家们通常并不总能预知合成最终的结果。这种方法就是"组合化学"。

制备化合物的传统方法是怀着明确的目标，一次只生成一种化合物。在组合化学中，几百甚至几千种化合物被同时生成，然后测验它们是否具有合适的行为，有时会成群出现性质类似的分子，当有希望的候选分子出现时，就只分析它们，鉴定它们的身份，并将其用作未来研究的基础。

这一流程源于肽链合成，它们是类似蛋白质的短链分子，我接下来会以它们为例说明这一流程如何实现。同种方法被用于研制一系列其他类型的化合物，并极大地提高了"药物研发"能力，而药物研发是研制具有药物活性的化合物的过程。

假设有三种氨基酸，A、B和C。第一轮中，我们准备一个含有A的容器，再与三种酸反应，得到AA、AB和AC三种化合物。然后将三种化合物放在一起，混合均匀，分成三等份，每一份都含有这三种化合物。第一轮反应就此结束。第二轮反应重复这一过程，与A反应的那个容器中会生成三种化合物AAA、ABA

和ACA,与B反应的第二个容器中会生成AAB、ABB和ACB,类似地,在第三个容器中会生成ACA、ACB和ACC。这九种化合物再被用作下一轮反应的起始反应物。在实际操作中,与此例中仅用于说明的三种化合物不同,自然界中所有20种氨基酸都会用到,连续四轮反应会产生4 008 000、160 000和5 200 000种化合物。所以,几乎眨眼间就可以(由机器人)制成数百万种化合物。

就这样,化学家们制成了各种物质,但他们并不总是清楚或关心制得了什么产物,而是期待在众多产物中找到珍珠。显然,这需要大量的记录工作从而了解每种混合物中可能存在的物质(例如,在三种氨基酸的例子中,第二轮后,第一个反应容器内只有三种候选分子),这时计算机就可以记录下来机器人的进展。如果在之后的测试中,一个混合容器内的物质显示出某种生物活性,例如可以抑制因某种酶失去活性而导致的某种疾病,那么这些物质就可作为分离与鉴定的候选分子,其他的就被丢弃了。

几年前,曾有一个令化学家们引以为傲的里程碑,他们生成并鉴定出了大约1 000万种化合物。如今,他们可以在一个月的时间内制得几倍于这个数目的化合物,只是偶尔才会费心去鉴定他们制得了什么。这就是进步。

第六章

成　就

　　我已经说过,没有化学的生活会让我们重回石器时代。几乎所有现代世界的基础设施与舒适生活都源于化学研究。在这一学科的原始时期,好奇心、传统及炼金术是它的搭档,科学仍旧处于萌芽状态,研究得不到正确理论的指导,进展缓慢。现如今这一学科已经成熟,好奇心积极有效地与理解、探索同行,研究在很大程度上是理性的,并取得了巨大成就。

　　最宽泛地说,化学家们发现了如何将某种形式的物质转变为另一种。在一些情况下,他们发现了怎样从地球上获取原材料,例如石油和矿石,并直接制成石油燃料和可用于制造钢的铁。他们也发现了如何从天空获取资源,从大气中分离出氮气,并将其转变成化肥。他们也找到了制备高度复杂的物质的方法,用作织物或当今我们认为的高科技所需的物质,他们知道未来还有更先进的科技。我们可以确信,化学将使这些成为可能。

土、气、火和水

我将从著名的古代四种元素——土、气、火和水——展开讲述化学的众多成就。

首先，在个体生物层面与全球社会层面上，水都是使得生命成为可能的绝对必要因素。化学通过净化水并除去其中的病原体，使得群居生活成为可能。氯是城市得以存在的主要因素：没了它，疾病就会蔓延，城市生活将变为一场赌博，更像是在城市中死亡，就像很早之前一样。化学家们找到了从氯含量丰富的源头——常见的食盐氯化钠——获取氯元素的方法，即通过电解熔融状态的盐氧化其中的氯离子，再从每个氯离子中剥离一个电子，将它们转化为氯元素。强力的氯气可以攻击病原体，使它们无害化。

在从咸淡水、地下蓄水层中有毒的水，以及拥有最丰富水资源的大海中获取饮用水这一领域，化学家们站在解决难题的最前列。他们发明了"反渗透"技术，从而直接帮助解决这项难题，在此项技术中，水被挤压通过渗透膜从而去除使水不可饮用的离子。他们也间接地帮助解决这项难题，他们发明了可抗高压的渗透膜，提高了过滤过程的效率。更不必说，化学家的经典分析技能对这一领域的探索至关重要，他们能够发现水中存在哪些物质、哪些物质可以容忍、哪些物质必须去除。

下面讲讲土，它是食物的来源。随着全球人口增长，肥沃土地面积减少，培育多产、高产作物变得愈发重要。基因工程（一项真正做到与生物学积极有效配合的化学技术）是探索方向之一，但由于各种原因仍然备受争议，一些原因听似有理，其他的则不然。增加农作物产量的传统方法是使用化肥。在此，化学家们为寻找经济可行的氮、磷来源，及确保它们可被转化为植物可以吸收的形式，做出了巨大贡献。

空气可为土壤供给原料。大气中富含氮（N），一种农业必需的元素，它占大气总量的近四分之三；但是它存在的形式无法被大部分植物吸收。它难以改变的惰性几乎完全归咎于氮分子（N_2）中的两个氮原子以一个强大的三键结合在一起，众所周知，这三对共用电子极难分开。事实上，这也是大气中富含氮的主要原因：它对大多数尝试与之进行反应的努力都保持冷漠，需要闪电或某些豆科植物的细菌以化学方式将其固定。

在20世纪的最初几年，化学最大的贡献之一就是发现如何从空气中获取氮，并将其转化成可被农作物吸收的形式（还被用于制作炸药），这一成就起初并不是出于为人类提供食物的人道目的，而是出于杀戮的非人道目标。由弗里茨·哈伯和卡尔·博施取得的这一项成就，是化学工业的一个里程碑，因为除了需要找到合适的催化剂用于加速以氮气与氢气制成氨（NH_3）的反应，它还要求工厂进一步发展，能在前所未有的温度、压力下进行生产。今天全球仍在使用这种方法生产，它还是需要消耗大量能量的。可以将苜蓿、三叶草、菜豆、豌豆以及其他荚豆类根部寄生的根瘤菌内发生的反应扩展到工业规模并用于收集大气中的氮，这是我们梦寐以求的愿望。化学家们已经在这一可能性上投入了几十年的研究，仔细研究了细菌内的酶如何以安静、节能、低压、低温的方式实现固氮。目前成功的曙光隐现，但还没有找到商业上可行的方法。

自然界中同样富含磷（P），它们来自史前动物的遗骸。这些动物由磷酸钙组成的骨头，以及它们体内特殊的能量来源——ATP（三磷酸腺苷）分子为我们与它们体内的每个细胞提供能量——都以磷酸盐岩的形式堆积、压缩、埋藏于世界海洋之下。在此，化学家们找到了从埋藏的资源中提取磷并再次用于可持续的大循环的方法，他们为开采死物来养活活物做出了贡献。

在了解过水、气及可培育出食物的土之后，我们还需要了解代表火的能量。没有能量，世界上什么都不会发生，如果不能再获得能量，文明就会崩塌。文明的进步得益于不断使用更多的能量，而在开发新能源、更有效利用已有能源方面，化学家们在各个层面、各个方面做出了贡献。

当然，汽油是一种极其便利的能量来源，因为它可以被方便地运输，甚至可以使用对重量敏感的飞机运输。化学家们长久致力于精炼从地下挤出与抽取出的原材料。他们已开发出了不同的流程与催化剂，可以将从自然界获取的分子切分成更易挥发的片段，并重塑它们，使其更高效地燃烧。但在后代眼中，焚烧自然界给予的地下馈赠可能是对珍贵资源的大肆破坏，类似于物种灭绝。石油也是有限的，尽管到目前为止还在不断发现经济上可行的新石油资源，但同时也证实了对其进行开采十分危险、成本在逐渐增加。我们必须意识到，尽管挖空地球在几十年后才会发生，但这一天终究会来到，我们需要有预期。

化学家们到哪里寻找新的能量来源呢？太阳，这个天空中遥远、剧烈的核聚变熔炉，就是一个明显的能量来源，而光合作用是一种自然界采用的摄取太阳能量的方法，显然，这种方法是一个可以效仿的模式。化学家们已经发明了效率适中的光伏材料，还在持续提高它们的效率。大自然早在40亿年前就开始了实验化学家的工作，发展出了一个以叶绿素为基础的高效系统，尽管已理解整个过程的主要特征，化学家仍需要解决一个难点，就是利用大自然提供的模式，并将其改造以用于工业规模。一种方法是利用阳光将水分解成组成它的元素，即想得到的氢气和已经十分充足的氧气，然后将氢气输送或泵送到可以燃烧利用它的地方。

我提到了"燃烧"。通常，燃烧后可收集到以热的方式释放出的能量，用于机械的、低效率的发动机或发电机，化学家们知道除了这种方法外，还有对以氢和烃类为代表的这类能源更精妙、更有效的使用方式。**电化学**利用化学反应产生电力，再以电力促进化学变化，这种方法可能对世界具有深远意义。化学家们已经在生产移动电源，即电池方面发挥了重要作用，电池用于驱动我们身边的各类小型移动设备，例如电灯、音乐播放器、笔记本电脑、电话、各种监控设备，并逐渐应用于机动车。

在发展"燃料电池"方面，化学家们在所有规模上都与工程师们携手前行，从为笔记本电脑供电再到为整栋房屋，甚至为整个村庄供电。在燃料电池中，从外部供应的氢或烃类燃料，可以通过化学反应在导电界面上倾倒和提取电子，从而产生电。燃料电池可行性的关键取决于发生反应的界面和浸没界面的介质。

即便是利用核能，无论是核裂变还是未来某一天的核聚变（在地球上模拟太阳中的反应），都需要依靠化学家的技术实现。建造核反应堆取决于新材料的可获得性，而将核燃料以铀及其氧化物的形式从它的矿物中提取出来需要化学。每个人都知道，除了政治和经济因素，阻碍核能的发展和公众接受度的因素在于，如何排放具有强烈放射性的核废料。化学家们的贡献在于找到了从核废料中提取可用同位素的方法,以及保证它不进入环境、成为世纪灾难的方法。

石油制品

上节中提到,我们将在地下埋藏了几亿年的复杂有机混合物——石油——从地下开采出来、随意燃烧,这一行为看似是在对无价资源的恣意破坏。当然,并不是所有的石油都用于发动机,其燃烧的产物从轿车、卡车、火车和飞机的排气管内喷出。其中很大一部分被提炼出来,用作化学家们发现的令人惊叹的连锁反应的源头,这就构成了石油化学工业。

环顾你的身边,看一看化学家们所取得的成就,他们从地下开采出黑色的、黏稠的原油,让它进行他们研发出来的反应,再将得到的产物送给现代世界的人工制品的制造商。

可能这些生产流程所带来的最大影响就是塑料的发展。一个世纪前,日常世界由金属、瓷器或是天然材料构成,人们使用由木头、羊毛、棉花及丝绸制成的物品。如今,大量物品都是用由石油衍生出的合成物质制成的。我们使用的织物是由化学家们发明的材料织成的,旅行中所携带的行李袋和箱子也是由合成材料制成的;我们使用的电子设备,电视、电话及笔记本电脑都是由合成材料制成的。由合成材料制成的机动车逐渐增多。就连世界的样子和感觉今天都与一百年前大不相同:如今,你触碰某件物品,它的质感通常与合成材料类似。对于这一转变,我们需要感谢化学家,他们发现了如何将从地球溢出的长链分子切分开,再通过聚合反应将其重新组合成长链分子。所以,乙烯($CH_2{=}CHX$,其中$X{=}H$)被制成聚乙烯,并用于从购物(例如用作塑料袋,一项利弊共存的发明)到帮助赢得第二次世界大战的胜利(例如用于雷达电缆的保护层)的各个方面。就像我在第四章中提到的,当X是氯(Cl)的时候,这种分子的单体可被制成聚氯乙烯,它在很多建筑工程中取代了木头和金属。

尽管塑料袋给环境带来的灾难可能超过它带来的便利,但是试想一下,如果没有由化学家们设计出来,随后又大量生产的聚合材料,我们会失去哪些东西。试想一下,一个没有尼龙和聚酯面料以用于服装、室内装潢以及装饰的世界是什么样。试想一下,一个只能用沉重的金属器皿盛装饮品、食物及家用液体的世界是什么样。试想一下,一个日常生活中没有小塑料制品的世界是什么样,它们包括开关、插头、插座、玩具、刀柄、键盘、按钮……这份清单几乎无穷无尽,以化学方法制成的聚合材料就是如此无处不在。

即使你对很多天然材料的逝去感到痛心,你同样可以感谢化学家,是他们使其在仍然使用之处得到保存。天然物质会腐坏,但是化学家们发明了可以延迟腐坏的材料。简言之,化学家们既可以提供合适的或所需的新材料,也可以在经过判断而选择使用天然材料的情况下,提供延长其寿命的方法。

塑料仅代表过去一百年材料革命的一个方面,至今这场革命仍在活跃地持续进行。化学家们发明了开始取代机动车内金属的陶瓷材料,如此可以减轻它们的重量、提高运输系统的效率,从而减轻对环境造成的负担。当然,陶瓷材料有悠久历史,它们被制成碗罐(另一项对社会生活可行性的未得到足够认可的贡献)。现代陶瓷由经过纯化的黏土及其他材料更系统地制成,有时会表现出出人意料的特性。例如,谁会想到用由近乎女巫的方式挑选出的元素所制成的陶瓷会展现出惊人的**超导**特性，即无电阻导电的能力？这种材料要在很低的温度下使用，但已远高于之前已知的超导材料的使用温度，所以在经济上更令人振奋、可以接受，然而我们仍在探索这种材料的应用领域，因为制造陶瓷电缆和陶瓷膜仍是一个具有挑战性的课题。

玻璃材料可归类于陶瓷。现代玻璃包括光缆，它构成了全球通信系统的支柱。玻璃本质上是沙子里的硅石（二氧化硅，SiO_2），它被提纯，熔融，再冷却。几个世纪以来，化学家们摆弄这一基本成分，带给我们颜色丰富的"彩色"玻璃，其迷人的色调是由精心地、选择性地加入杂质形成的。当然在早期，这些颜色依靠的是玻璃匠人的技艺与智慧，而不专靠化学家的。但如今是由化学家们制定玻璃成分的，在有些情况下玻璃展现出丰富的颜色，在另一些情况下，例如光缆，则是完全透明的，可以将光脉冲传导至极远的距离之外，只伴随极少量的衰减。

创造色彩

没有化学家们的贡献，人类创造出的世界将会很乏味。鲜艳的颜色曾只属于有钱人，他们承担得起天然色素的开销，比如从某些海螺（染料骨螺）的腺体黏液中提取出泰尔紫，这一过程需要挤压12 000只海螺从而获取一克多的染料，而这仅够为长袍的褶边上色，或是为了得到深邃迷人的群青色，他们会购买从遥远的阿富汗运来的青金石（"天堂石"）。之后，为了让帝国的军队和官员免于疟疾，威廉·珀金（1838—1907）在不了解奎宁结构的情况下尝试合成它，但最终没有成功，他误打误撞地合成了一种他称为苯胺紫的染料，于是他没能拯救患病的士兵，却成功地拯救了海螺，还顺带创立了英国化学工业。由此他为积累自己的财产与英国的大部分资产奠定了基础。

化学家为物质世界添加了一整个色彩光谱，使世界不再乏味，除非有这个需求（例如用于伪装），除此之外，物质世界可以是明亮果敢的也可以是端庄含蓄的。如今，颜色不仅范围宽广，比如荧光、闪光已列入此范围，而且非常牢固，可以经受强力的洗衣过程。

化学创造的色彩并不局限于布料。颜料普遍得到发展；不仅上色材料本身，支持介质也发展了，例如用于楼房的颜料与艺术家使用的丙烯颜料。想一想家用颜料的进步，例如改进了颜料的流动特性，提高了在严苛环境下的稳定性，扩大了颜色的范围，其中包括那些特意做出用后变淡效果的颜色。

即便是电视机屏幕与电脑显示器的颜色也利用了化学家发明的固体。使用既耗电又笨重的阴极射线管的日子已一去不复返。如今，我们身处液晶、等离子显示和OLED（有机发光二极管）的世界。液晶与OLED都由化学家研制的分子构成，这些分子可以以特殊方式回应电场，让可视化便携设备成为可能。

日常生活中的基础设施

化学家们同样为研发半导体做出了贡献，它们是现代世界的通信与计算的基础。确实，如今化学最主要的贡献之一就是改进了构建电子世界的物质基础设施。化学家们研制出了用于计算与光导纤维的关键材料——半导体，它们正逐渐取代铜在信号传输中的重要地位。显示屏作为与人类视觉系统的连接界面，也是化学家们在材料研发方面取得的成果。

如今，化学家们正在研发分子计算机，其中的开关与存储器都基于分子形状的改变。这些材料的成功研发——鉴于对科学一贯抱有的乐观态度，我们可以相信这份努力终将通向成功——会带来前所未有的算力提高与令人吃惊的紧凑性。如果你对这类智慧材料的发展感兴趣，那么你就可以期待将有一场计算领域的变革。除此之外，还有量子计算方面的发展前景，它取决于化学家研制出适用的新材料，这将会在通信与计算领域引起几乎难以预想的重大变革。

药物化学

我还几乎没有提及健康。化学对人类文明（必须提到，还包括兽群福利）最大的一个贡献就是药物的研发。化学家们有理由为研发出对抗疾病的药剂感到骄傲。可能他们最受欢迎的贡献就是研发了麻醉剂从而减轻了预期疼痛。回想二百年前，人们只能用喝白兰地和咬牙坚持的方式接受截肢手术！化学家们的下一项重要贡献在于发现了抗生素，通常化学家们会仔细观察自然，从中找到研发方向。一百年前，细菌感染的预期结果只有死亡，但是现在有了盘尼西林及其经过化学改良的后代，感染可被治愈。我们只能期待这些药物可以继续发挥作用，但同时也要为相反的情况做好准备，因为细菌会进化，可以规避它们的天敌。

制药公司经常因为人们认为它们获取了巨额利润、剥削他人而受到指责。但在深思后，他们值得同情。他们的内在动机来自一个值得称赞的目的，即发明战胜疾病的药物从而减轻人类的痛苦。化学家就处于这项事业的核心。不幸的是，研发成本极高。现代计算技术可以帮助寻找新方法、减少对活体动物实验的依赖,但是将外源材料引入人体时需要极其小心,因为如果在最后测试阶段发现不可接受的后果,那么几年的高额经费研究就很可能会功亏一篑。

化学家们在减轻疾病领域做出的贡献,还与他们在分子水平上的参与密切相关。半个多世纪前,当DNA结构被发现的时候(1953年),生物学就成了化学。在很大程度上,分子生物学就起源于这一发现,它是研究有机体功能的化学。化学家们经常伪装成分子生物学家,他们在非常基础的层面上打开了通往理解生命及其主要特点——遗传——的大门,从而让分子世界中更广阔的领域进入了理性研究的视野。他们也改观了法医学,将犯罪分子绳之以法,同时还改观了人类学。

化学将关注点投向生命过程的时机正好处于传统化学的分支——有机化学、无机化学、物理化学——发展到相当成熟的阶段,并且准备好应对有机体,尤其是人体内运行的极为复杂的过程网络之时。化学家们持续的新发现为疾病治疗,更为重要的是,为疾病预防奠定了合理的基础。如果你计划进入这个领域,那么基因组学与蛋白质组学将对你的研究极其重要。这是一个真正可以让你对化学建立起信心的领域,使你感受到站在早先巨人的肩膀上,同时坚信你在破解疾病的根源。

战争以及其他恶果

化学也有黑暗的一面。只谈及化学的丰功伟绩,而闭口不谈它曾被用于增强人类破坏与杀戮的能力是有失公允的,因为这些伟大成就的背后有巨大的代价,有些关乎人命,另一些则关乎环境。

第一，杀戮与残害能力提升了。化学家们需要对研发战争中使用的毒气和改良炸药负责。事实上，我之前提到的弗里茨·哈伯，他除了发明了合成氨工艺、使强效化肥得以广泛使用外，同样引领了毒气的研发。尽管我们会指责他的品性，但还是有人希望，禁止这类武器会使我们更友善地评价他给人类生活带来的净贡献。即便政府对使用这类可怕的武器负有责任，但在我看来，为研制这类武器做出贡献的化学家们同样难逃其责。无论从哪一角度进行评判，研发化学武器都绝对有百害而无一利，无法减轻我们的谴责：它们就是纯粹的罪恶。许多国家，并非都是最强大的国家，但占世界人口约98%的国家都已经将它们当作非法武器而禁止使用，希望剩下的几个国家可以追随这一主流、加入禁止这类武器的协议。

化学战也可以在无意中发生。例如1984年发生在印度博帕尔的事件，联合碳化物公司位于这一地区的工厂失控，官方资料显示，这直接造成将近4 000人死亡，两周内又有8 000人死亡，超过50万人受伤。即便是有预谋的化学战都从未造成过如此严重的伤亡。灾难的直接原因是水进入了一个储量过多、冷却不足的异氰酸甲酯（CH_3NCO）储藏罐，异氰酸甲酯是一种杀虫剂生产过程中的中间体。那段期间，对这种杀虫剂的需求下降了，这导致了中间体储量过多。水如何进入储藏罐仍有待商榷：公司坚称是某个心怀不满的员工造成的人为破坏；其他人坚称由于工厂安全管控混乱、无效、缺失、不当以及不受重视，水偶然进入了储藏罐。之后发生的反应向大气中释放了30吨有毒气体，造成了附近人口密集的贫民窟内的居民大量死亡，并经受身心上无法估量的痛苦。

对化工厂固有的危害进行评论是多余的,其中危险大于益处的观点是陈词滥调。这些灾难只发生在极少数情况下,我们只能期盼从沉痛的代价中汲取教训,逐步改进工厂的设计和运行,从而在总体上增进我们的福祉。

化学的另一个黑暗面是对炸药的供应、改良以及生产。这一面并不完全是黑暗的,因为炸药在采石和开矿方面发挥着重要作用。黑暗的部分在于将它们制成炸弹,以及为发射子弹、迫击炮等提供推动力。炸药是一类引爆后可快速发生反应的化合物——在原理上,分子碎裂成小块,形成气体,气体的极速生成会产生爆炸时破坏性的、脉冲式的冲击。

在炸药历史的早期阶段,火药曾是王者。它的爆炸取决于氧化剂(硫、硝酸钾)与可被氧化物质(木炭,本质上是一种碳元素的不纯形式)的充分混合。电子从碳转移到氧化剂,它们拖拽着原子,这产生了大量的小分子,即气体。从那时起,人们研制出了反应更快,因而冲击力也更强的物质和混合物。化学家们尝试了替代混合不同物质的反应方式,要实现极限充分的反应:确保氧化和可被氧化的部分都出自同一分子,让电子转移和随后的原子重排、分子碎裂可以尽快发生,并形成大量可增强冲击效果的小碎片分子。最著名的这类分子就是硝酸甘油。阿尔弗雷德·诺贝尔(1833—1896)驯服了这种极其不稳定的化合物,他发现这种化合物可被吸进一种多孔黏土,制成炸药。不久后,硝酸甘油为最可抚慰良心的基金会之一,诺贝尔基金会的设立提供了资金,它的奖项都致力于改善人类的生存状况与宣扬和平。

环境问题

既然我们处在谈论化学负面影响的尴尬角落,就不得不提到另一项严重的指责,即它,至少它的排放物应为环境破坏负责。无法否认化工厂排放的废水给生态环境带来了巨大灾难。自从珀金的工厂依照每日生产重点将附近的运河染成红色、绿色、黄色,人类所向往的更加美好的生活就一直以牺牲环境为代价。事实上,环境污染的苗头(如果这种说法不算过于讽刺)可以追溯到希腊、罗马时期,因为冰芯分析显示了那一时期的金属制造活动留下的痕迹。

在这条前进的路上,可以使用法律或化学方式解决问题。法律通过惩罚预期来限制污染,化学通过从源头上消除来避免污染。后者总是更优的处理方式,它取决于化学自身的发展,还促成了一场政治-环境-化学三方面协调的**绿色化学**运动。从广义上讲,绿色化学通过建立有关材料使用与废物消除的严格规章制度,最大限度地减少化学生产过程对环境造成的影响。

绿色化学的支持者们首先提出了一个貌似合理的主张,即预防废物产生优于事后清理。这项基本原则的含义是,生产过程中的任何起始原料都应该完整地出现(尽可能接近地)在最终产物中:无论反应中加入什么原子都应该出现在产物的分子中,尽可能少丢弃不需要的原子。这一含义对经济与科技领域都产生了巨大冲击,因而商业上也不愿意采用,因为生产过程与工厂需要按规定设计,特定原材料难以获取,这可能会造成成本高昂。

过程优化后,尤其是如果优化程度已超出技术与经济的范畴，那么这些反应应该设计为能避免或至少降低有毒化合物成为废料或可逃逸中间体的可能性。这项要求同样适用于最终产物，它应该带给人类生活（这项要求明确指明是人类，但加入其他生物更加合适）和环境最低的毒性风险。这条限制也适用于过程中使用的辅助材料，尤其是用作溶剂的液体，它们可能，也许在一些当下使用的工序中由"可能"变为"一定"，会因回收时发生泄漏，即使是少量的泄漏，而排放到环境中。化学家们是必不可少的嗅探者，即便在他们的小型实验反应中，也可以选出良性的溶剂并发现在这些不常见的新条件下发生的反应。

绿色化学的支持者们的另一个理想的愿望就是使用可再生的反应原料。可再生性有很多形式，但所有形式都要避免从地球内部开采资源。大自然每年都会为我们供应农作物，由于这个过程源于太阳的恩惠，它以光合作用为媒介推动二氧化碳循环，所以农作物算作可再生原料。除二氧化碳之外，其他物质也可循环，已经有将垃圾填埋场用作矿场的提案，但是这种资源极其危险，超出地质学的评判范围。

绿色化学的支持者们也意识到了另一个产生废物和污染的因素：在化学反应中能量所扮演的角色。一切对能量的需求都诉诸环境，或是以需求燃料的方式或是以排放废气的方式对环境产生影响。在理想情况下，所有反应都应该以不加热不冷却的方式进行，况且冷却更昂贵、破坏性更强。

此外，还有许多技术性更强的要求，使生产过程尽可能绿色环保。有机化学中的许多反应，例如在制药领域，都需要多个中间步骤，分子在成为最终产物前会被暂时修饰。每一步反应都需要在特定条件下进行，需要专门的化学试剂，也许还会用到各种有毒性的溶剂。通过减少中间产物、寻找更直接地从原料到产物的反应途径，反应可以移向生产谱系的环保端。

除了生产过程本身，聪明的绿色化学家们还进一步关注产品的整个使用期限，他们寻找各种手段，保证在临近产品使用寿命的时候，它自身及其分解形成的任何物质都不具有毒性，也不会在环境中降解为有害残留物。"整个使用期限"的考量包括在生产过程中（回想博帕尔事件）预见会出现的灾难，其防范性的意义在于，无论生产或是存储的物质是什么，都应该在事故发生时对环境产生最低影响。降低灾难的可能性依赖持续、可靠的分析，分析对象包括反应与存储容器的所有部分与条件，并制定防故障监测程序，这些程序不能像在博帕尔事件中那样被忽略或跳过。

这些都是绿色化学向往的目标。其中深层次的考量在于以化学方式解决，最好是尽量避免它可能导致的问题。当然，追逐商业利润与承担社会、环境责任之间总会产生冲突，这种紧张关系不会在某些低水平监管的环境中得到缓和，这种低水平监管让工业得以逃脱对谋杀，真正意义上的谋杀的惩罚。

潘多拉魔盒总是如此：干预大自然总要承担风险。化学家们对物质自然的根源进行干预，用她提供的原子重新制成她不熟悉的化合物，这些化合物闯入她的生态系统，可能扰乱生命的微妙平衡。拥有魔法师梅林那样的对原子的操纵能力就意味着要承担起责任，在过去这种责任尚未被意识到，但在社会压力下，如今的化学工业对其自身该承担的责任有了充分的认识。

然而，关键问题在于，若不继续发展化学，那么如何有把握地解决世界上的各种难题。化学掌握了可改善我们日常生活的几乎每一方面的钥匙,由摇篮至坟墓以及其中的所有瞬间。它已经为我们的舒适生活提供了物质基础,不仅限于健康方面,还有疾病方面,而且没有理由假设它已发展到了顶峰。它为我们虚拟的和现实的通信做出了贡献,因为电子与光子在复杂的模式和相互作用的网络中移动,进而产生了计算,而化学为电子与光子的移动提供了介质材料。除此之外,它还改善了我们使用的燃料,让它们可以更高效地燃烧,并以催化方式减少了燃烧后的有害产物,它还帮助我们从使用化石燃料转向使用可再生资源,例如对光伏材料的研发。无论对土壤、空气,还是对水来说,化学都是解决由它引发的环境问题的唯一方法。

化学在文化方面的贡献

化学的另一项成就也是我们在研究中不该忽视的:它让我们深刻理解物质世界的运作方式,理解范围从石头直至有机体。这种理解改善了人类的状况,因为它解释了好奇心,从而给我们增添了乐趣。

通过化学,我们了解了构成地貌的矿物的成分和结构,明白了石头的结构,从而知道了它们为什么坚固、为什么会闪闪发光、为什么会碎裂风化,以及它们含有什么成分。我们知道了为什么金属可被打造成不同形状、拉成丝,通过对金属原子排布的了解,知道了为什么有一些可以按照我们的意愿弯曲,但是另一些会断裂。我们了解了通过形成合金和钢可以发挥不同金属的特性。我们了解了宝石的颜色,还有我们为什么可以透过玻璃而不能透过木头看到东西。

通过化学,我们可以阐释、理解自然界中曾被认为神秘莫测的现象。我们可以理解叶子的绿色与玫瑰的红色,还可以理解香草与新割草料的芬芳。我们可以虽不时停顿,却越发清晰地理解自然界中精密复杂的过程之网,它构成了令人敬畏而又多面的特性,我们称之为生命。我们开始理解我们的大脑内发生的各种化学过程,它赋予了我们认知、好奇以及理解的能力,而这一理解的停顿之处更多。

尽管化学不研究物质世界的终极构造,即处于基本粒子物理学领域中的基本粒子动物园,但是化学研究这些基本粒子的组合,即原子,这些原子都具有鲜明的个性。通过化学,我们得以了解不同元素的特性,就是它们为什么通过原子结构而具有这些特性,以及它们为什么只能形成特定的组合,而非别的组合。通过化学,化学的根本,我们知道了如何利用这些特性来构建分子和物质形式,而它们在银河系的其他地方可能根本不存在。

通过化学,我们了解了食物的味道、织物的颜色、物质的质感、水的潮湿、春夏秋季叶子颜色的转变。并不是在生活中的每时每刻我们都需要使用理解力,因为以动物那种愉快的方式躺下,单纯地沐浴在身边的喜悦之中本身也是一种快乐。但是化学为这种喜悦增加了深度,因为我们在情绪高涨、心血来潮之时,可以看透世界的表面愉悦,而享受对事物本质的认知。

第七章

未 来

新元素还在不停地被发现，如今的速度是每年一种左右，这意味着理论上元素周期表越来越大，从而给化学家们更广阔的探索空间。不幸的是，这些新元素没什么用：它们具有放射性、极其不稳定，在几分之一秒内就会消失。除此之外，至今为止只合成了这些元素的几个原子，它们转瞬就会消失，变为基本粒子。

未知的边缘

出于理论方面的原因，化学家们怀疑在元素周期表中接下来的部分，在还没合成的数个126号前后的元素那里（2013年我们发现了116号元素鉝，还有其他几个尚未命名的元素，虽然还没有亲眼见到，但有微弱证据显示它们存在）会形成一个"稳定岛"，其中的元素比之外的元素可以存在更久。不过，除了作为研究原子核结构理论的实验台外，它们几乎没有任何其他可用之处。化学家们也没理由认为它们会为化学的发展提供动力。

化学家们手中有很多可使用的元素。很多新方法也正在开发中,有望提升观测的灵敏度、准确性及广度。可探测出极少量物质是福也是祸。了解样品中的详细成分可加深我们的理解,检测出恐怖分子手中炸弹的蛛丝马迹可帮我们生存下去,但在这个持续动荡的世界中,随处都可以找到污染物只会为我们带来困惑,并可能使我们过于警觉。

新世界

处于开发中的重要技术包括那些可直接研究少量原子和分子的技术,它们将取代通过观测大量样品来推知原子和分子行为的方法。化学家们想知道分子间相互作用与转变的亲密关系,而可以检测到单个分子的性质,或它们结合到一起发生反应——化学键松动、原子挣脱限制而重新排列——时的性质是化学(至少是物理化学)梦寐以求的圣杯。近些年来,我们可以观测水分子在飞秒(10^{-15}秒,1/1 000 000 000 000 000秒)量级的时间尺度上的逐步变化,并且在将时间尺度缩小至阿秒(10^{-18}秒,比之前快一千倍)方面取得了进展,在这一尺度上,即便是电子的运动也会定格,最终化学成了物理学。

当我们遇到微小的原子团时,有趣的问题和特殊的规则会发生作用。以水为例:最小的冰块有多大?研究表明,你至少需要275个水分子才能使分子团表现出类似冰的性质,大约475个水分子才会使分子团真正成为冰。这是一个立方体,每条边上有八个水分子。这类知识的重要性在于,它有助于我们模拟大气中云的形成过程,并理解液体如何凝固。

当研究低温下的一小组原子的时候,我们需要接受它们的行为由量子力学支配,并预料到它们会有不寻常的特性。所有物质，包括日常物质，都由量子力学支配，但是我们面对的是数量庞大的原子，因此即使是一小撮盐，其不寻常的特性也会被冲淡，我们只能感知到平均状态，即我们所熟知的普通物质的行为。这些人们开始制造出的具有新状态的物质可能对化学本身并不重要，但是也说不定：它们可能在存储数据、发展量子计算方面成为绝佳选择。

化学家们正为一个研究少量分子的新兴领域做出巨大贡献：纳米科学与纳米科技的世界。纳米系统（源于希腊文 nanos，小矮人）由直径大约为100纳米（10^{-7}米，1/10 000厘米）的实体组成，它处于单个分子（比它小1 000倍）与大块物质（比它大1 000倍）的中间区域。前沿领域总是令人着迷，这个介于大与小之间的概念前沿领域也不例外。纳米粒子（注意"纳米"这一前缀可以加到很多名词前面，未来将会更多）小到可以表现出量子效应，小到让人们对曾被当作完备理论的热力学感到困惑并重新审视它。

对物理化学家而言，这是一个成果丰硕的探索领域，他们可以构想、完善传统理论，从而用于这些未曾有过的非传统材料。在这一领域内，有机和无机化学家也有很多可贡献之处，尤其在制备纳米材料方面，因为研制出的有机和无机物质都可以用于纳米领域。制备过程可以"自上而下"，从宏观材料雕刻出纳米材料，就像雕塑师在大理石上雕刻；也可以"自下而上"，一砖一瓦地建造出纳米结构。后者尤其有趣，因为其建造通常以"自组装"形式进行。在这个不需要人工干预的过程中，分子被摇晃混合在一起，就会聚合成想要的纳米结构，就像我们曾经可能幻想过的，随便摇晃拼图碎片，它们就可以自动扣在一起而组成一幅图片,并不用麻烦地动手将它们拼在一起。

如今,纳米技术,即纳米材料的开发和应用,与纳米科学,即对它们的整体研究,在化学领域蔚然成风,这是理所当然的,因为纳米材料有远大的前景。所有研究机构都投入了这类研究。纳米材料的潜在应用领域横跨几个学科,并已成为众多实际应用的核心。例如,与传统硅太阳能电池相比,它们展现出更优的光捕获特性,已用于血糖传感器。由于担心有毒的镉元素可能不适于注射进人体内,所以人们已对用于血糖传感器的含镉材料进行了广泛研究;但最近在灵长类动物身上的实验结果减轻了这方面的忧虑。纳米棒、纳米线、纳米纤维、纳米晶须、纳米带及纳米管也都研制出来了,有望用于纳米机械与纳米计算机领域。

化学即将在计算微型化方面发挥重要作用。我们已见证了缩小计算机体积(以及降低能耗)产生的影响,从20世纪50年代可以填满整个房间的早期计算机演变为如今微型的、无处不在的、功能强大的计算机,还见证了它们给社会和日常生活带来的影响。这是由米到厘米尺度的转变,计算机的线性尺寸缩小了100倍,体积与重量减少了100万倍,从房间大小缩小到口袋大小,但计算能力却大幅提升了。若现今分子计算领域的发展取得成果,那么之前曾发生过的尺寸缩小、能力提升、社会影响力随之增大的情况就可以再现。

计算过程取决于两方面:存储与处理。存储在分子层面很容易实现,只要引起分子形状的改变,这种改变可被保留下来并且通过某种观测方式获取。例如,可以使某个分子弯曲成某种形状,对应数字1,也可以弯曲成另一种形状,对应数字0。如今,可以进行多种构象变化，例如一个环状分子滑向一个棒状分子的某一端并停留在那里。处理更加困难，但核心就是从特定的输入得到特定的输出。化学是以化学反应的形式从输入得到输出，包括两种试剂相遇而产生光。

自然在研发DNA的时候就已经解决了数据存储问题，也已经进化出提取这些信息并将其转化为有机体的方法。我们的记忆在大脑中以未知的方式进行化学编码，提供了一个巨大却又脆弱且存储不完善的数据库。DNA分子被用于进行简单的算数运算，如果它们遇到了一个破损的蛋白质分子，就会"决定"采取必要的治疗方式。培养计算机而非制造计算器仍是科学幻想，但是有迹象表明这一幻想即将成为现实。

新维度

化学研究从三维转向二维是最近值得注意的发展趋势。常用来填充铅笔的材料石墨是碳元素的一种形式，其中的碳原子会形成类似铁丝网的扁平薄片，当存在杂质的时候，这些薄片会划过彼此，可能在纸上留下痕迹或用作润滑剂。单个的薄片被称为**石墨烯**，可以以一种简单的方法将其从固体石墨上剥离下来，这一事实使安德烈·海姆与康斯坦丁·诺沃肖洛夫获得了2010年的诺贝尔物理学奖。

现今，石墨烯本身被视为物理学家，可能还有工程师的巨大财富。它是已知强度最高的材料之一，其断裂点比钢高200倍，但它质量极轻，每平方米不到一克。诺贝尔奖颁奖辞评价道："一平方米的吊床可以支撑一只四公斤的猫，但重量只有一根猫胡须。"它无与伦比的电学、热学及光学特性也令我们非常感兴趣，其潜在的应用包括：制造无可动部件的喇叭并模压到不同的表面上，以及实现在室温下蒸馏伏特加，本质上是除去其中的水。

化学家们对这笔宝贵的二维材料财富研究到了什么程度？它如今被用于实验室技术，例如用作分离不同类型分子的筛子（目标之一为生产生物燃料）以及用于海水淡化（使海水可饮用）。尽管石墨烯本身不易吸附气体分子，它的表面——几乎全是表面——可以通过化学修饰对不同类型的气体分子作出反应，而气体分子的附着可以改变石墨烯薄片的电学特性，因而这些分子的存在可以被检测到。

化学家们当然会想知道这种二维材料的仙境能否为其他材料居住，并测试这些材料能否避免石墨烯的美中不足之处。通过电化学反应方法，我们已经制得具有石墨烯形式的新材料，例如硫化钼、硫化钨，以及以碳化钛为基础的更奇特的材料。在这些二维材料之中，有一些具有石墨烯缺少的半导体特性，它们被用于微型集成电路。石墨烯本身易于进行化学修饰，其中一种操作是将其氧化为氧化石墨烯。这种材料的薄片可以聚合形成"石墨烯纸"，材料科学家们希望它可以成为一类具有可调节电学、热学、光学及机械特性的新材料的基础。

新应用

化学家与材料科学家、物理学家、生物学家以及工程师合作研发出的新材料应用极为广泛，我所做的无异于站在阿拉丁的神秘洞穴中朝周围随意指点，我知道我会遗漏重要的发展或例子，但还是希望可以通过几个例子来传达这样的印象：他们的这种合作正在改变我们的生活。

所以,我想以自洁玻璃为例。这项省力的发明基于光化学以及对分子间吸引力、排斥力,尤其是表面"疏水"或憎水特性的理解。典型的自洁玻璃涂有一层薄且透明的二氧化钛,在阳光的作用下,可通过化学方式分解附着在其表面上的污物。憎水表面意味着,任何水,特别是雨水,都可以清洗经光催化分解形成的产物,而不留污痕。

我还可以以智能织物为例。智能织物可以发出不同颜色的光,可能呈现穿戴者身上的温度分布,进而粗略地反应出他们的情绪状态。或者它们可以根据周边环境或是穿戴者的突发奇想,以电学方式改变外观。这些织物不仅要有趣,还必须经得起洗涤,经得起穿着、褶皱、折痕带来的应力。

我已讲过催化对工业极为重要,此外它对消除内燃机的污染也必不可少。如今,所有的汽车都装有催化转化器,它利用了一些极其复杂的化学反应,因为它不仅要在发动机冷启动后(那时会产生大量污染)迅速开始工作,还要在发动机热得烫手时继续工作。除此之外,催化剂不仅要将氮氧化物还原为无害的氮气,还要将一氧化碳氧化为二氧化碳,并将未充分燃烧的烃类燃料完全氧化。不仅如此,它们还需要对发动机工作时的不同条件作出反应,例如在油气混合物不足或过剩的情况下,以及在突然加速的情况下。所有这些都需要化学家的研发工作。

可能现代化学最重要的用途就在于,研发可以对抗疾病、缓解疼痛、提升生命体验的新药物。**基因组学**涉及基因鉴定以及基因在管理蛋白质生产的过程中的复杂的相互作用,它对现今和未来的药物基因组学的发展至关重要。药物基因组学研究基因信息如何改变个体对药物的反应,这为个体化医疗,即根据个人的基因组成量身定制药物的鸡尾酒,提供了前景。

比基因组学更复杂的是**蛋白质组学**,即对组成某个有机体的所有蛋白质的研究,这些蛋白质在生命层面上发挥作用,也是大多数药物在体内的作用点。在这一领域中,计算化学必须联合医学化学,因为如果可以确定导致某种疾病的蛋白质,并想终止它的作用,那么对可能侵入并阻断蛋白质活性位点的分子进行计算机建模,就是合理研发药物的第一步。这也是通向高效、实用的个体化医疗的另一条途径。

新发现

我不想让你觉得,化学取得的进展仅限于与之相关的应用领域。它们诚然是吸引人眼球的头条新闻并与我们所有人息息相关。然而,化学家们还从事基础研究,深入了解物质以及它们如何被修饰。他们逐渐熟悉自然在分子层面上的运作,学习她的方式,偶然发现一些可能令人惊叹的特性,这些特性除了成为最宝贵的知识之外,可能暂时看来并没有任何其他可用之处。基础研究在这方面绝对起着举足轻重的作用,因为它给我们带来意想不到的发现、意想不到的理解,以及意想不到的智慧非凡的应用。

为了画上一个圆满的句号,我想在此提及一个仅有的、奇特的、独特的、纯粹学术性的新发现:化学家们发现,大自然本身具有打结的能力。科研人员惊喜地发现了一类可自发折成三叶结的分子。就像一位评论员所评述的,"这项新研究诠释了合成化学与物理有机化学最精妙的那些方面，而且它也是为数不多的对立体化学极致精美的呈现"。

这些就是现代化学所唤起的快乐，思维的乐趣。我期待这些文字可以在某种程度上抹去那些曾污染了你对这门非凡学科的看法的记忆，还希望你能分享这份快乐。

第七章 未来

元素周期表

族	1	2											13	14	15	16	17	18
																		2 He 氦
周期 1	1 H 氢																	
2	3 Li 锂	4 Be 铍			3	4	5	6	7	8	9	10	11	12	5 B 硼	6 C 碳	7 N 氮	8 O 氧	9 F 氟	10 Ne 氖
3	11 Na 钠	12 Mg 镁											13 Al 铝	14 Si 硅	15 P 磷	16 S 硫	17 Cl 氯	18 Ar 氩
4	19 K 钾	20 Ca 钙	21 Sc 钪	22 Ti 钛	23 V 钒	24 Cr 铬	25 Mn 锰	26 Fe 铁	27 Co 钴	28 Ni 镍	29 Cu 铜	30 Zn 锌	31 Ga 镓	32 Ge 锗	33 As 砷	34 Se 硒	35 Br 溴	36 Kr 氪
5	37 Rb 铷	38 Sr 锶	39 Y 钇	40 Zr 锆	41 Nb 铌	42 Mo 钼	43 Tc 锝	44 Ru 钌	45 Rh 铑	46 Pd 钯	47 Ag 银	48 Cd 镉	49 In 铟	50 Sn 锡	51 Sb 锑	52 Te 碲	53 I 碘	54 Xe 氙
6	55 Cs 铯	56 Ba 钡	57 La 镧	72 Hf 铪	73 Ta 钽	74 W 钨	75 Re 铼	76 Os 锇	77 Ir 铱	78 Pt 铂	79 Au 金	80 Hg 汞	81 Tl 铊	82 Pb 铅	83 Bi 铋	84 Po 钋	85 At 砹	86 Rn 氡
7	87 Fr 钫	88 Ra 镭	89 Ac 锕	104 Rf 𬬻	105 Db 𬭊	106 Sg 𬭳	107 Bh 𬭛	108 Hs 𬭶	109 Mt 鿏	110 Ds 𫟼	111 Rg 𬬭	112 Cn 鿔	113	114 Fv 𫓧	115	116 Lv 𫟷	117	118

镧系 6 | 58 Ce 铈 | 59 Pr 镨 | 60 Nd 钕 | 61 Pm 钷 | 62 Sm 钐 | 63 Eu 铕 | 64 Gd 钆 | 65 Tb 铽 | 66 Dy 镝 | 67 Ho 钬 | 68 Er 铒 | 69 Tm 铥 | 70 Yb 镱 | 71 Lu 镥 |

锕系 7 | 90 Th 钍 | 91 Pa 镤 | 92 U 铀 | 93 Np 镎 | 94 Pu 钚 | 95 Am 镅 | 96 Cm 锔 | 97 Bk 锫 | 98 Cf 锎 | 99 Es 锿 | 100 Fm 镄 | 101 Md 钔 | 102 No 锘 | 103 Lr 铹 |

词汇表

酸 质子的供体（参看 路易斯酸）。

碱液 水溶性碱；碱在水中的溶液。

氨基酸 化学式为 $NH_2CHRCOOH$ 的有机化合物（R 代表原子团，例如—CH_3，或更复杂的结构）。

分析 对物质的鉴别及其量与浓度的测定。

阴离子 带负电荷的原子或原子团。

原子 某种元素的最小粒子；电子围绕原子核所构成的实体。

碱 质子的受体（参看 路易斯碱）。

键 两个原子之间共用的一对电子。

碳水化合物 典型化学式为 $(CH_2O)_n$ 的有机化合物。

催化 由不发生净反应的化学物种加快化学反应。

阳离子 带正电荷的原子或原子团。

链式反应 一个分子、离子或自由基攻击另一个，产物继续攻击下一个，以此类推的连续反应。

络合物 由中心金属原子及与其结合的配体形成的原子团。

化合物　元素通过化学键形成的特定结合体。

衍射　波在传播路径上遇到障碍物时产生的干涉现象。

双键　两个原子之间的两对共用电子。

电化学　利用化学反应产生电能，以及利用电能带来化学变化。

电解　通过施加电流实现化学反应。

电子　带负电荷的亚原子粒子。

亲电体　被电子密集（带负电的）区域吸引的化学物种。

亲电取代　反应物之一为亲电体的取代反应。

元素　无法通过化学方法分解为更简单物质的物质；由单一种类原子组成的物质。有关元素及其符号的列表，参看前文的元素周期表。

基因组学　基因的鉴定以及基因在管理蛋白质生产的过程中的复杂的相互作用。

绿色化学　通过严格指导材料的使用与废物的消除，最大限度地减少化学生产过程对环境的影响。

水合氢离子　H_3O^+。

氢氧根离子　OH^-。

中间体　参看 **反应中间体**。

离子　带电的原子或原子团（参看 **阳离子**与**阴离子**）。

同位素　原子核的原子序数（质子数）相同，但中子数不同的原子。

路易斯酸　电子对受体。

路易斯碱　电子对供体。

路易斯酸碱反应　路易斯酸与路易斯碱发生的 $A + :B \rightarrow A—B$ 形式的反应。

配体　络合物中与中心金属原子结合的原子团。

孤对电子　不直接参与成键的一对电子。

混合物　不形成新化学键的物质混合。

分子　化合物中的最小粒子；以确定的方式排列的原子的离散结合体。

单体　聚合反应中使用的小分子。

亲核体　被缺电子（带正电的）区域吸引的化学物种。

亲核取代　反应物之一为亲核体的取代反应。

氧化　使物质失去电子；与氧气的反应。

光子　一种电磁辐射粒子。

聚合物　聚合反应的产物。

聚合　小分子连结在一起形成长链。

产物　由化学反应得到的物质。

蛋白质　由氨基酸组成的复杂化合物。

蛋白质组学　对组成某个有机体的所有蛋白质的研究。

质子　氢原子的原子核。

自由基　有至少一个未成对电子的化学物种。

反应物　特定化学反应中的起始原料。

反应中间体　除反应物与产物外，参与反应机理的物质。

反应试剂　在各种化学反应中用作反应物的物质。

氧化还原　一种化学物种被氧化、另一种化学物种被还原的反应；电子转移反应。

还原　一种化学物种得到电子的反应。

盐　酸与碱反应生成的离子化合物。

溶质　被溶解的化学物质。

化学物种　这里指代原子、分子或离子。

光谱学　对样品吸收辐射或发射辐射的观测。

取代反应　对分子中存在的原子或原子团进行取代的反应。

超导性　无电阻导电的能力。

合成　由较简单的物质创造出新物质。

滴定　通过测量中和一种碱液（或酸液）所需的酸液（或碱液）的体积,来确定该碱液（或酸液）的浓度。

过渡金属　元素周期表中第3至第11族的成员。

索 引

（条目后的数字为原书页码，见本书边码）

A

Absorption spectroscopy 吸收光谱学 56

Acetic acid 乙酸 40

Acid 酸 40

Activation barrier 活化能垒 34

Activation energy 活化能 33

Adenosine triphosphate 三磷酸腺苷 69

Adsorb 吸附 54

AFM 原子力显微镜技术 61

Air 空气 67

Alchemy 炼金术士 1

Alfalfa 苜蓿 68

Alkali 碱液 40—41

Aluminium oxide 氧化铝 46

Amino acid 氨基酸 63

Ammonia 氨 23, 35, 68

Amputation 截肢手术 76

Analysis 分析 53

Analytical chemistry 分析化学 10

Anion 阴离子 20—21

Antibiotics 抗生素 76

Atom transfer 原子转移 46

Atomic force microscopy 原子力显微镜技术 61

Atomic number 原子序数 15

Atomic spectroscopy 原子光谱学 55

Atomic structure 原子结构 14

Atoms 原子 2

ATP 三磷酸腺苷 69

Attosecond timescale 阿秒量级 87

B

Balance 天平 1

Base 碱 41

Basid reaction 酸碱反应 45

Battery 电池 46

Bhopal 博帕尔 78

Biochemistry 生物化学 10

Biofuel 生物燃料 91

Biology as chemistry 生物学作为化学的延伸 5

Bolinus brandaris 染料骨螺 74

Bond 键 24

Bond formation 键的形成 19

Bosch, C. 卡尔·博施 36, 68

Bottom-up 自下而上 88

Bragg, W. and L. 威廉·布拉格与劳伦斯·布拉格 59

Breathing 呼吸 50

Brønsted, J. 约翰内斯·布仑斯惕 40

Burette 滴定管 53—54

Burning 燃烧 69

C

C_{60} C_{60} 62

Carbon 碳 8

Carbon dioxide 二氧化碳 29

Carbon monoxide poisoning 一氧化

化学

碳中毒 50
Carbon valence 碳价 23
Carrot and cart 胡萝卜加小拖车 36
Catalysis 催化 92
Catalyst 催化剂 34—35
Catalytic converter 催化转化器 92
Cation 阳离子 20
Ceramics 陶瓷 73
Chain reaction 链式反应 48
Chemical equilibrium 化学平衡 35
Chemical kinetics 化学动力学 32
Chemical weapons 化学武器 78
Chemistry, structure of 化学的结构 7
Chlorine 氯气 67
Chlorophyll 叶绿体 70
Chromatography 色谱法 54
Classical mechanics 经典力学 4
Clay 黏土 73
Cloud (of electrons) (电子)云 17
Cloud layers 云层 18
Combinatorial chemistry 组合化学 64
Combustion 燃烧 30
Common salt 食盐 21, 41, 67
Complex 络合物 49—50
Computational chemistry 计算化学 62, 63
Cooking 烹饪 34
Corrosion 腐蚀 43
Covalent bonding 共价成键 22
Cyanide poisoning 氰化物中毒 50

D

Dalton, J. 约翰·道尔顿 2

Deuterium 氘 16
Diffraction 衍射 60
Disease 疾病 76
Disorder 混乱度 28
Distillation 蒸馏 54
DNA data storage DNA数据储存 90
Double bond 双键 24
Driving power 推动力 28
Drug development 药物研发 93
Ductile 可延展的 25
Dye 染料 50
Dynamic equilibrium 动态平衡 36
Dynamite 炸药 79

E

Earth 土 66, 67
Electric conductivity 导电性 25
Electrochemistry 电化学 70
Electrode 电极 46
Electrolysis 电解 46
Electron 电子 14—16, 43
Electron cloud 电子云 17
Electron pair 电子对 24
Electron sea 电子海洋 25
Electron shells 电子层 17—18
Electron spin 电子自旋 24
Electron transfer 电子转移 43, 79
Electrophilic substitution 亲电取代 52
Element 元素 2—3
Elements of antiquity 古代元素 67
Elements, order of 元素排序 15
Emission spectroscopy 发射光谱 55

Endothermic reaction 吸热反应 31
Energy 能量 4—5, 27
Enthalpy 焓 30
Entropy 熵 28
Entropy increase 熵增 31
Enzyme 酶 35
Equilibrium 平衡 35
Ethylene 乙烯 72
Exothermic reaction 放热反应 31
Explosives 炸药 78

F

Fabric 织物 72
Faraday, M. 迈克尔·法拉第 43
Femtosecond timescale 飞秒量级 87
Fertilizer 化肥 66
Filtration 过滤 54
Fire 火 67
Fire-retardant 阻燃剂 48
Fireflies 萤火虫 34
First Law of thermodynamics 热力学第一定律 27
Fission 核裂变 71
Forensic chemistry 法医化学 10
Franklin, R. 罗莎琳德·富兰克林 59
Free radical 自由基 47
Fuel cell 燃料电池 70
Fusion 核聚变 71

G

Geim, A. 安德烈·海姆 90
Genomics 基因组学 93

Glass 玻璃 73
Graphene 石墨烯 90
Graphical representation 图像呈现 63
Graphite 石墨 90
Green chemistry 绿色化学 11, 80
Gunpowder 火药 79

H

Haber, F. 弗里茨·哈伯 36, 68, 78
Haber-Bosch process 哈伯-博施法 35
Haemoglobin 血红蛋白 50
Health 健康 76
Heat 热 30
Heavy hydrogen 重氢 16
Heavy water 重水 16
Hodgkin, D. 多萝西·霍奇金 59
Homeostasis 内环境稳态 36
Hydrochloric acid 盐酸 40
Hydronium ion 水合氢离子 42
Hydrophobic 疏水的 92
Hydroxide ion 氢氧根离子 41

I

Industrial chemistry 工业化学 11
Infrared radiation 红外辐射 56
Infrared spectroscopy 红外光谱学 56
Inorganic chemistry 无机化学 9
Insight 理解 83
Interference 干涉 60
Ion 离子 20—21
Ionic bonding 离子成键 22
Isotope 同位素 16

K

Knots（绳）结 94

L

Laboratory equipment 实验仪器 39
Lapis lazuli 青金石 74
Lead-acid battery 铅蓄电池 46
Leguminous plants 豆科植物 68
Lewis acid 路易斯酸 49
Lewis acid-base reaction 路易斯酸碱反应 49
Lewis base 路易斯碱 50
Life as titration 与滴定过程类似的生命 42
Lifetime span 使用期限 82
Ligand 配体 49
Liquid crystal 液晶 75
Lithium-ion battery 锂离子电池 46
Livermorium 鉝 3
Lowry, T. 托马斯·劳里 40
Lustrous 有金属光泽的 25

M

Macroscopic world 宏观世界 3
Magnesium burning 镁燃烧 43—44
Magnesium chloride 氯化镁 44
Magnesium oxide 氧化镁 44
Malleable 可塑的 25
Mass spectrometer 质谱仪 58
Mauveine 苯胺紫 74
Mechanism of reaction 反应机理 33

Medicinal chemistry 药物化学 12
Mendeleev, D. 德米特里·门捷列夫 13
Metal 金属 24—25
Metallic bonding 金属成键 25
Methane 甲烷 23
Methyl isocyanate 异氰酸甲酯 78
Microscopic world 微观世界 3
Molecular biology 分子生物学 11—12, 77
Molecular computer 分子计算机 75

N

Nanobelt 纳米带 89
Nanofibre 纳米纤维 89
Nanomaterial 纳米材料 88, 89
Nanoparticle 纳米颗粒 88
Nanorod 纳米棒 89
Nanosoccer 纳米足球 62
Nanosystem 纳米系统 88
Nanotechnology 纳米科技 88
Nanowhisker 纳米晶须 89
Nanowire 纳米线 89
Neutron 中子 15,16
Newton, I. 艾萨克·牛顿 4
Nitrogen 氮 14, 66, 68
Nitrogen fixation 固氮 68, 69
Nitroglycerin 硝酸甘油 79
NMR 核磁共振 56
NMR spectrum 核磁共振谱 57
Nobel Foundation 诺贝尔基金会 80
Nobel, A. 阿尔弗雷德·诺贝尔 79
Non-metal 非金属 26

Novoselov, K. 康斯坦丁·诺沃肖洛夫 90
Nuclear atom 原子核模型 14
Nuclear magnetic resonance 核磁共振 56
Nuclear spin 核自旋 57
Nuclear waste 核废料 71
Nucleophilic substitution 亲核取代 52
Nucleus 原子核 14—15

O

Oil 石油 71
OLED 有机发光二极管 75
Optical fibre 光缆 73
Ore reduction 矿石还原 44
Organic chemistry 有机化学 8
Organic light-emitting diode 有机发光二极管 75
Organism 有机体 5
Organometallic chemistry 金属有机化学 9
Oxidation 氧化 43, 44
Oxidation defined 氧化的定义 44
Oxygen 氧 14

P

Pathogens 病原体 67
Peptide 肽链 64
Periodic Table 元素周期表 13, 86
　structure of 其构造 18
Perkin, W. 威廉·珀金 74

Personalized medicine 个体化医疗 93
Pesticide 杀虫剂 78
Petrochemicals 石油化学制品 69
Petroleum 汽油 69
Pharmaceutical companies 制药公司 76
Phosphorus 磷 14, 69
Photon 光子 55, 56
Photosynthesis 光合作用 47, 70
Physical chemistry 物理化学 7—8
Pigments 颜料 50, 74
Pipette 移液管 53—54
Plastic bags 塑料袋 72
Plastics 塑料 48, 71
Poison gas 毒气 78
Polyester 聚酯 72
Polyethylene 聚乙烯 48, 72
Polymerization 聚合反应 48
Polystyrene 聚苯乙烯 48
Polytetrafluoroethylene 聚四氟乙烯 48
Polythene 聚乙烯 48
Polyvinyl chloride 聚氯乙烯 48
Pots 碗罐 73
Powder diffraction pattern 粉末衍射模式 60
Preservation 保存 72
Product 产物 38, 39
Protection 保护 51
Protein folding 蛋白质折叠 63
Proteomics 蛋白质组学 93
Proton 质子 15, 39—40
Proton spin 质子自旋 57
Proton transfer 质子转移 39—40

PTFE 聚四氟乙烯 48
PVC 聚氯乙烯 48, 72

Q

Quantum mechanics 量子力学 4
Quinine 奎宁 74

R

Radical 自由基 47—48
Radioactivity 放射性 16
Rate of reaction 反应速率 32, 33
Reactant 反应物 38
Reaction mechanism 反应机理 33
Redox reaction 氧化还原反应 45—46
Reduction 还原 44, 45
Rontgen, W. 威廉·伦琴 59
Rutherford, E. 欧内斯特·卢瑟福 15

S

Salt 盐 41
Scanning tunnelling microscopy 扫描隧道电子显微镜技术 61
Sea-snail 海螺 74
Seawater 海水 91
Second Law of thermodynamics 热力学第二定律 5, 28
Self-assembly 自组装 88
Self-cleaning glass 自洁玻璃 92
Semiconductors 半导体 75
Separation 分离 54
Shaking and stirring 晃动与搅拌 38

Silica 硅石 73
Single bond 单键 24
Smart fabrics 智能织物 92
Soap 肥皂 40—41
Sodium 钠 21
Sodium chloride 氯化钠 21, 41, 67
Sodium sulfate 硫酸钠 41
Solid-state chemistry 固体化学 9
Solvents 溶剂 81
Spectroscopy 光谱学 8, 55
Spin 自旋 24
Stained glass 彩色玻璃 73
STM 扫描隧道电子显微镜技术 61, 62
Subatomic particle 亚原子粒子 3
Substitution reaction 取代反应 51
Sulfur 硫 14
Sulfuric acid 硫酸 40
Superconductivity 超导性 73
Surfaces 表面 61
Synthesis 合成 53

T

Teflon 特氟龙 48
Texture 质感 72
Thermochemistry 热化学 31
Thermodynamics 热力学 4—5, 27
Thomson, J. J. 约瑟夫·约翰·汤姆逊 43
Titanium dioxide 二氧化钛 92
Titration 滴定 53—54
Top-down 自上而下 88
Toxicity 毒性 81
Transition metal complex 过渡金属

化学

络合物 49
Triple bond 三键 24
Tritium 氚 16
Two-dimensional chemistry 二维材料化学 91
Tyrian purple 泰尔紫 74

U

Ultraviolet radiation 紫外辐射 56
Union Carbide 联合碳化物公司 78
UV-vis spectroscopy 紫外-可见分光光度法 56

V

Valence [化合] 价 29—30

Vibrational spectroscopy 振动光谱学 56
Vitalism 活力论 9

W

Water as a base 水作为一种碱 42
Water as an acid 水作为一种酸 42
Water molecule 水分子 22, 29
Watson, J. 詹姆斯·沃森 59
Weighing 称量……的重量 2
Wilkins, M. 莫里斯·威尔金斯 59

X

X-ray diffraction X 射线衍射法 59
X-rays X 射线 59—60

Peter Atkins

CHEMISTRY

A Very Short Introduction

Contents

    Preface i

1   Its origins, scope, and organization 1

2   Its principles: atoms and molecules 13

3   Its principles: energy and entropy 27

4   Its reactions 38

5   Its techniques 53

6   Its achievements 66

7   Its future 86

    Periodic table 95

    Glossary 97

    Further reading 101

Preface

I hope to open your eyes and show you a fascinating, intellectually and economically important world, that of chemistry. Chemistry, I have to admit, has an unhappy reputation. People remember it from their schooldays as a subject that was largely incomprehensible, fact-rich but understanding-poor, smelly, and so far removed from the real world of events and pleasures that there seemed little point in coming to terms with its grubby concepts, spells, recipes, and rules. In later life that unhappy reputation is often rendered unhappier still by an awareness of the environmental impact of nasty chemicals escaping into the wild and bringing disaster to softly green clover-clad bucolic meadows that were home to the glowing poppy and the dancing butterfly, rendering into inhospitable mud the banks where the wild thyme once grew, generating toxic sludge and noxious slime where limpid streams had rippled, replacing air fragrant with aeolian delight with pungency, and generally messing things up.

I want to change all that. I want to encourage you to look anew at chemistry, through modern unprejudiced eyes, with those memories and attitudes swept away and replaced by comprehension and appreciation. I want to show you the world through a chemist's eyes, to understand its central concepts, and see how a chemist contributes not only to our material comfort but also to human culture. I want to explain how chemists think and how what they reveal about matter—all forms of matter, from rocks to humans—adds pleasure to our perception of the world. I want to show you how chemists take one form of matter, perhaps sucked or dug from the ground or plucked from the skies, and turn it into another form, perhaps to clothe us, feed us, or comfort us.

I want to share with you the thought that chemistry provides the infrastructure of the modern world. There is hardly an item of everyday life that is not furnished by it or based on the materials it has created. Take away chemistry and its functional arm the chemical industry and you take away the metals and other materials of construction, the semiconductors of computation and communication, the fuels of heating, power generation, and transport, the fabrics of clothing and furnishings, and the artificial pigments of our blazingly colourful world. Take away its contributions to agriculture and you let people die, for the industry provides the fertilizers and pesticides that enable dwindling lands to support rising populations. Take away its pharmaceutical wing and you allow pain through the elimination of anaesthetics and deny people the prospect of recovery by the elimination of medicines. Imagine a world where there are no products of chemistry (including pure water): you are back before the Bronze Age, into the Stone Age: no metals, no fuels except wood, no fabrics except pelts, no medicines except herbs, no methods of computation except with your fingers, and very little food.

Advances in technology demand the availability of materials with new and sophisticated properties, be it better electrical, magnetic, optical, or mechanical properties or just greater purity. Advances in the maintenance of human health that can reduce the demand on the physical infrastructure of hospitals and their sophisticated, expensive equipment depend on the discovery and manufacture of better, more sophisticated medicines. There will be no advances in the generation, deployment, and conservation of energy without chemistry to provide its material infrastructure.

It goes without saying, however, that the extraordinary difference between raw nature and what chemistry transforms it into to enhance and extend our lives comes at a price, and it is that price that disconcerts us and is rightly the basis of our apprehension of chemistry's environmental impact. At its crudest, the products of chemistry enhance our ability to kill and maim, for weaponry is improved when new explosives and other agents are perfected. Often of more permanent and vocal concern is the undeniable environmental impact of what is produced and the processes of production. Chemistry puts into societies' hands the ability by governmental choice to wage war more effectively, through commercial pressures to produce artefacts more aggressively, and through personal choice to squander more profligately and thereby harm our unique and irreplaceable ecosystem.

I shall confront that concern in these pages, for it has been a corollary of progress in chemical manufacturing and the presence not only of its products but also of its manufacturing waste in the environment. It is important, though, to bear in mind a rounded picture of chemistry, not a single black facet. Without chemistry life would be nasty, brutish, and short. With chemistry, it can be comfortable, entertaining, and well fed. Transport can be efficient; clothes alluring. Lives can be longer. Without ignoring the dark and negative side of chemistry, I shall encourage you to appreciate the illuminating and positive side, too.

There is another dimension to all these contributions: understanding. Chemistry provides insight into the heart of matter by showing how things are. A chemist can look on a rose and understand why it is red and look on a leaf and understand why it is green. A chemist can look on glass and understand why it is brittle and look on a fabric and understand why it is supple. The glories of Nature, of course, can be experienced without this inner knowledge, just as music can be enjoyed without analysis; but the insight that chemistry brings into the properties of matter, in all its forms, can be brought to bear if the moment is apt, and deeper enjoyment thereby achieved. I seek to share some of this insight with you and show that even a little chemistry will add to your daily pleasure.

That, in broad terms, is the journey I shall take you on. I shall try to dislodge you from your half-remembered, perhaps unpleasant memories of your early encounter with chemistry. You will not have a degree in chemistry when you have read through these chapters, for chemistry is deep as well as wide, it is quantitative as well as qualitative, it is subtle as well as superficial. You will, however, I hope, appreciate its structure, its core concepts, and its contributions to culture, pleasure, economy, and the world.

In conclusion, I would like to thank Professor David Phillips, Imperial College, for a number of helpful remarks.

Peter Atkins
Oxford, 2014

Chapter 1
Its origins, scope, and organization

Greed. Greed inspired humanity to embark on an extraordinary journey that touches everyone today. The particular variety of greed I have in mind was jointly the quest for immortality and the attainment of unbounded riches. The supposed route to both was the manipulation of matter to provide elixirs to overcome bodily ills for the realization of immortality and recipes for the conversion of more or less anything resembling gold—either in colour, as in urine and sand, or heft, as in lead—into gold itself. Neither aim was ever achieved, but the ceaseless tinkering with matter by the alchemists provided them with a considerable familiarity with it and provided the compost, often literally, from which a real science—chemistry—was to emerge.

The principal instrument of the transition from alchemy to chemistry was the balance. The ability to weigh things precisely put into humanity's hands the potential to attach numbers to matter. The significance of that achievement should not go by unremarked, for it is in fact quite extraordinary that meaningful numbers can be attached to air, water, gold, and every other kind of matter. Thus, through the attachment of numbers, the study of matter and the transformations that it can undergo (the current scope of chemistry) was brought into the domain of the physical sciences, where qualitative concepts can be rendered quantitatively and tested rigorously against the theories that surround and illuminate them.

Weighing matter before and after it had undergone transformation from one substance to another led to the principal concept that underlies all explanations in chemistry: the *atom*. The concept of the 'atom' had floated around groundlessly in human consciousness for over two millennia, ever since the ancient Greeks had speculated, without an iota of evidence, for some kind of ultimate indivisible particulate graininess of the world. Their speculation became grounded in science in the hands of John Dalton (1766–1844), who through the analysis of the weights of substances before and after reaction drew the conclusion that the elements, the fundamental building blocks of matter, are composed of unchangeable atoms, and that track could be kept of them as one substance changed into another by the simple expedient of weighing.

Atoms are now the currency of chemistry. Almost every explanation in chemistry refers to them, either as individuals or strung together in the combinations we call *molecules*. Atoms are the constituents of all matter: everything you can see and touch is built of atoms. As small as they are it is quite wrong to say that they are invisible to the naked eye. Look at a tree: you are seeing atoms. Look at a chair: you are seeing atoms. Look at this page: you are seeing atoms (even if this page is on a screen). Touch your face: you are touching atoms. Touch a fabric: you are touching atoms. Of course, an individual atom is too small to see: but matter is built from battalions of them, and the swarming battalions are visible to the naked eye as the substances that surround us. Later, however, in Chapter 5, I shall explain how chemists can now even see images of *individual* atoms.

There are just over 100 different types of atom. Quite what I mean by 'type' I shall explain in Chapter 2 when together we look inside them and identify their differing internal structures that render them distinct. Each different type of atom corresponds to a different element. Thus, just as there are the elements hydrogen, carbon, iron, and so on, so there are hydrogen atoms, carbon atoms, iron atoms, and so on, all the way up to the most recently discovered element, which in 2013 is the wholly useless and exceedingly short-lived 114th element, livermorium. (To be precise: it is element 116, but two that precede it await discovery). The key idea in chemistry is that when one substance changes into another, the atoms themselves do not change: they simply exchange partners or enter into new arrangements. Chemistry is all about divorce and remarriage.

Although 'atom' means uncuttable, atoms are cuttable. Even armchair speculation leads to that conclusion, for the existence of different types of atom implies the possession of different structures, so with sufficient ingenuity it is likely that an atom can be blasted apart and the so-called *subatomic particles* from which it is formed identified. Experiment confirms this speculation, and we shall see something of the interior of atoms and thereby the origins of their different personalities in Chapter 2. It is here that chemistry draws most strongly on physics, for physicists unravelled the structures of atoms and chemists use this information to account for the molecules they form and the reactions they undergo.

That last remark hints at the scope of chemistry. It implies that to understand chemistry it is necessary to import concepts from physics. That is indeed the case, and chemistry draws heavily on numerous concepts developed by physicists (in return, we chemists provide the matter for them to conjure with). Among all this trade there are two hugely important imports, one relating to the behaviour of individual atoms and their subatomic components and the other relating to bulk, that is tangibly large versions of matter, such as a jug of water or a block of iron. More technically, these are the *microscopic* and *macroscopic* worlds, respectively.

The crucial import from physics to account for the properties of the microscopic world of individual atoms and molecules is *quantum mechanics*. Although much of chemistry was developed during the 19th century, there was little understanding of why some things occurred and others did not. At that time, Isaac Newton's 'classical mechanics', the mathematical procedures for accounting for the motion of bodies, was king, for it was so successful at accounting for the orbits of planets and the flight of balls, and there was the expectation that when planets and balls were slimmed down to atoms, explanations of chemistry would be found and Newton's domain would encompass chemistry too. Newton's fruitless focus on alchemical manipulation was perhaps a sign that he thought so too. However, at the end of the 19th century and early in the 20th it was found that this slimming down of planets and balls to atoms resulted in the complete failure of classical mechanics: even the concepts on which Newton's mechanics was based crumbled when applied to atoms and their constituents. Such are the dangers of uncircumspect extrapolation.

Then, early in the 20th century, around 1927, a new mechanics was born that has proved to be hugely successful for explaining how atoms and subatomic particles go about their business. To this day the theory, quantum mechanics, has not been superseded in predictive power and numerical precision. That it remains largely incomprehensible is admittedly an irksome deficiency, but in due course I shall do my best to distil from it what is necessary for understanding the behaviour of atoms and hence the whole of chemistry. We shall see that when chemists stir and boil their fluids, they are coaxing atoms to behave according to the weird laws of quantum mechanics.

The other crucial import from physics, in this case to account for the properties of the macroscopic world of bulk matter, is *thermodynamics*. Thermodynamics is the science of energy and the transformations it can undergo. It arose in large part through the Victorian era's dependence on the steam engine for driving societies forward both literally and economically, but soon proved to be a key part of the fabric of chemistry. The material fabric of our subject is atoms, but the changes they undergo are under the control and impetus of energy. We shall see that not only is energy released when a fuel burns—an obvious, useful but primitive aspect of the involvement of energy with chemistry—but also that it governs how atoms behave in general, what structures they can form, what changes in organization they can undergo, and at what rate those changes can occur. Energy also, in a subtle way, turns out to be the driving power of chemistry in the sense that reactions are impelled forward by it in a manner that I shall explain in Chapter 3. Because energy is so intimately coiled into the very structure of chemistry, it should not be surprising that thermodynamics plays a role despite its engineering origins.

Whereas chemistry reaches down into physics for its explanations (and through physics further down into mathematics for its quantitative formulation), it reaches upwards into biology for many of its most extraordinary applications. That should not be surprising, for biology is merely an elaboration of chemistry. Before biologists explode in indignation at that remark, which might seem akin to claiming that sociology is an elaboration of particle physics, let me be precise. Organisms are built from atoms and molecules, and those structures are explained by chemistry. Organisms function, that is, are alive, by virtue of the complex network of reactions taking place within them, and those reactions are explained by chemistry. Organisms reproduce by making use of molecular structures and reactions, which are both a part of chemistry. Organisms respond to their environment, such as through olfaction and vision, by changes in molecular structure, and thus those responses—all our five or so senses—are elaborations of chemistry. Even that hypermacroscopic phenomenon, evolution and the origin of species, can be regarded as an elaborate working out of the consequences of the Second Law of thermodynamics, and is thus an aspect of chemistry. Some organisms, I have in mind principally human beings, cogitate on the nature of the world, and the mental processes that underlie and are manifest as these cogitations are due to elaborate networks of chemical reactions. Thus, biology is indeed an elaboration of chemistry. I shall not press the view, whatever I actually think, that all matters of interest to biologists, such as animal behaviour in general, are also merely elaborated chemistry, but confine myself to the assertion that all the structures, responses, and processes of organisms are chemical. Chemistry thus pervades biology, and has contributed immeasurably to our understanding of organisms.

We socially elaborate organisms, we humans, build things. We fabricate artefacts. We mine the stones of the Earth, pump the fluids from the deep, and harvest the gases of the skies and aim to turn all this raw material into whatever we desire. The conversion of those raw materials into substances that can be moulded, hammered, spun, glued together, eaten, or simply burned, is a part of chemistry. Chemists might step aside and allow moulders to mould, hammerers to hammer, shapers to shape, and in general fabricators to fabricate, to create the final artefact, but it is they who have provided the raw material, the infrastructure of our modern technological society, and have contributed hugely thereby to world economies and the deportment of individuals and nations.

As I have emphasized in the Preface, there are, of course, speckles and blotches of black amid all this light. Chemistry has certainly contributed to mankind's ability to maim and kill, and it would be inappropriate in this survey of what chemistry is to sweep under the carpet of its pages its provision of explosives, of nerve gases, and its accidental and intentional impositions on our fragile environment. I shall confront these issues later, but at this stage—to emphasize the importance of personal judgement—I invite you to eliminate all the contributions of chemistry to the modern world, which will take you back to the painful, dangerous, uncomfortable, aspirationally restricted era of the Stone Age, and to ask yourself then whether the current darkness outweighs the light.

The divisions of chemistry

The scope of chemistry, then, is so enormous that my introduction to it, and the subject itself, would wallow amorphously like a stranded spineless whale without the imposition of some kind of structure. Chemists have drifted into a structure that helps them to carry out their activities, congregate in like-minded assemblies, and develop their procedures much like individual states develop their policies and economies. Unlike most states, the boundaries are blurred, and often striking advances are made where two cultures overlap. That is especially the case when the subject is as mature as chemistry currently is, where each domain of activity is thoroughly explored and inspiration might most fruitfully come, just like in art, at fertile overlapping boundaries and at the frontiers where chemistry overlaps other disciplines.

For our purposes, and to understand the general structure of chemistry for the sake of this introduction, it is helpful to appreciate its division into various branches and to see in broad terms their concerns. The divisions of chemistry still pervade university departments, courses, and the journals where discoveries are reported, and so a description of them is still an important component of a visitor's guidebook. But be warned: frontiers both intellectual and departmental are melting.

The broadest, most important and conventional, and still widely observed division of chemistry is into its physical, organic, and inorganic branches.

Physical chemistry lies at the interface of physics and chemistry (hence its name) and deals with the principles of chemistry which, as we have seen, consist largely of quantum mechanics for explaining the structures of atoms and molecules and thermodynamics for assessing the role and deployment of energy. It is also concerned with the rates at which reactions take place, both at the macroscopic level and the microscopic. In the latter it seeks to follow the intimate lives of individual molecules as they are ripped apart and then reconstituted as different substances in reactions. A major activity of physical chemistry is its contribution to the interpretation of investigative techniques, particularly 'spectroscopy'.

As we shall see in Chapter 5, spectroscopy uses various kinds of light to bring information from within molecules into the eyes, increasingly the synthetic eyes, of the observer. Such is the current sophistication of these techniques that physical chemists must bring all their armoury, particularly quantum mechanics, to bear on the interpretation of the data. Indeed, so blurred are the activities of chemists and physics in this domain that the name physical chemistry often elides into *chemical physics* for some who study the behaviour of individual molecules with an approach that lies close to a physicist's.

Organic chemistry is the part of chemistry that is concerned with the compounds of carbon. That one element can command a whole division is a testament to carbon's pregnant mediocrity. Carbon lies at the midpoint of the Periodic Table, the chemist's map of chemical properties of the elements, and is largely indifferent to the liaisons it enters into. In particular, it is content to bond to itself. As a result of its mild and unaggressive character, it is able to form chains and rings of startling complexity. Startling complexity is exactly what organisms need if they are to be regarded as being alive, and thus the compounds of carbon are the structural and reactive infrastructure of life. So extensive are the compounds of carbon, currently numbering in the millions, that it is not surprising that a whole branch of chemistry has evolved for their study and has developed special techniques, systems of nomenclature, and attitudes.

Why 'organic'? Such is the intricacy of the molecules to which carbon contributes (except for a few outliers, like simple carbon dioxide), that it was originally thought that only Nature could form them. That is, according to this 'vitalist' view, they are the products of organisms. The beginning of the end of vitalism was in 1828, when it was shown that a simple mineral could be converted into a characteristic 'organic' compound (namely urea). Although dispute raged for some time, since then the 'organic' of organic chemistry has been an archaism; but convenient archaisms are hard to dislodge and the term survives but now means nothing more than 'a compound of carbon'.

That leaves the rest of the elements, the hundred or so elements other than carbon. Their study is the domain of *inorganic chemistry*. As might be suspected about a branch of a subject that deals with over 100 elements with widely differing personalities, inorganic chemistry is a vital yet sprawling field of study. The sprawl is partly contained by the adoption of various subdivisions of the subject. A major subdivision is *solid-state chemistry*, where the object of study is inorganic solids, such as the materials that act as superconductors and the semiconductors that have made universal computation feasible. It is hard to resist the analogy between inorganic chemistry and a hundred-piece orchestra, with the chemist conductor–composer drawing out symphonies of combinations by ordering the instruments accordingly.

Carbon is not secure from an inorganic chemist's Periodic-Table-scanning eyes. Some of the simpler compounds of carbon, such as the carbon dioxide that I have already mentioned, the killer gas carbon monoxide, and chalk and limestone that form our landscapes, are readily released by organic chemists from their domain as being of little interest to them and by convention are regarded as inorganic. On the frontier between the divisions, though, lie compounds that are intricate assemblies of carbon atoms yet include atoms of various metals. A number of these compounds are essential catalysts in the chemical industry; some are crucial to the functioning of organisms. Here lies the interdivisional field of *organometallic chemistry*, which at its best represents a highly fruitful collaboration between organic and inorganic chemists.

Chemistry's overlap with other disciplines

Such are the three principal divisions of chemistry. That list by no means exhausts all the ways in which chemists carve up their subject for better digestion, but all the others draw technique, concept, and inspiration from these three in various proportions and spice their mixture with aspects of other subjects. It would be a sizeable undertaking to list them all, but it is appropriate to be aware of the most common of them.

Analytical chemistry is the modern descendant of the age-old quest for finding out what is there. What is present in a mineral? Might there be gold or is it hafnium? What is present in crude oil? What is present in it other than the raw hydrocarbons, and which hydrocarbons? What is that compound you made? Can you deduce the arrangements of its atoms? These are all questions that analytical chemists might try to answer. Although test-tubes, flasks, and retorts still figure in their approaches, many of their investigations are now carried out in sophisticated machines, some of which use spectroscopy and others techniques developed by inorganic and physical chemists. I explore these techniques in Chapter 4. Stemming from analytical chemistry is *forensic chemistry*, in which the techniques of analytical chemistry are used for legal purposes, to track down or exonerate suspects, and to analyse the scenes of crimes.

Biochemistry is organic chemistry's back-donation to biology, sometimes with a dash of inorganic chemistry thrown in. It is concerned wholly with the structures and reactions that constitute living things, resolving the metabolic pathways that turn food into action (including that action confined to the brain: thought). Organisms are still a hugely important reservoir of organic molecules, for Nature has had billions of years to explore structural niches, and biochemists play a central role in both discovering what is there and working out how it was made under the control of the worker-bees of the body, the proteins we call enzymes. One anthropocentric but important concern about the extinction of species is that it wipes out sources of intricate molecules that have taken millions of years to emerge.

The name of *industrial chemistry* speaks for itself. Here chemist meets engineer, and reactions established in test-tubes and their kin are scaled up to enormous size and rendered fit to contribute to commerce. Industrial chemists contribute enormously to the economy and to trade between nations. In the United Kingdom alone, chemicals contribute 20 per cent to the gross domestic product, and in the United States over 96 per cent of all manufactured goods are directly touched by chemistry. Such figures relating to manufactured chemicals are almost literally not to be sniffed at. A principal concern of current industrial chemistry is *green chemistry*, the intention being to minimize waste, thereby enhancing economy, and to minimize impact on the environment, which enhances acceptability and sustainability.

The contribution of chemistry to other disciplines

Chemistry owes many debts to the subjects that surround it in the intellectual landscape, but they owe debts to chemistry too.

Physics owes debts, particularly in the field of electronics and increasingly of photonics (the use of light instead of electrons to convey information and manipulate data). Chemists create the semiconductors without which computation would be confined to the industrial scale from which it first emerged. They also formulate the glasses used in optical fibres, without which the transfer of information would be hobbled.

Biology owes an enormous debt to chemistry, especially since the emergence of *molecular biology*, springing largely from the identification of the structure of DNA and its interpretation as the carrier of genetic information from generation to generation. It is almost no exaggeration to say that biology became a part of the physical sciences once the chemical component of its principal characteristic, reproduction, had been identified. Molecular biology is really a version of chemistry, and the current maturity of chemistry has enabled biology to become as lively as it has never been before. The collaboration of biology and chemistry that we call *medicinal chemistry* is one of chemistry's great and unarguably acceptable contributions to society.

Society owes chemistry another huge debt too, for as I have said in the Preface, it deploys the material contributions of chemistry everywhere, in medicine, agriculture, communication, transport, and all forms of construction, fabrication, and decoration. We personally also owe a debt to chemistry, for as I claimed there too it gives us each an inner eye to enjoy the world.

All this stems from an understanding of chemistry, which I shall now start to unfold.

Chapter 2
Its principles: atoms and molecules

Central to any discussion of chemistry is the Periodic Table, that masterpiece of organization formulated in the 19th century principally by Dmitri Mendeleev (1834–1907) and its basis understood in the 20th century once the structures of atoms had been explained. That the table is important is confirmed by its ubiquity: it hangs on laboratory and lecture room walls and is printed in every introductory chemistry textbook. There is a version towards the end of this book. Its importance, though, should not be overstated. Working chemists do not gaze at it each morning for inspiration or refer to it frequently during the day. Certainly they have it in mind, for its real importance is that it summarizes relationships between the elements and plays a crucial role in organizing information about them. Perhaps its most important role is in the teaching of chemistry, for instead of being confronted with the daunting task of learning the properties of a hundred elements, it enables their properties to be inferred from their location in the table and trends in properties to be identified and easily remembered. Indeed, Mendeleev was led to formulate his table as he prepared to write a textbook of introductory chemistry.

The Periodic Table portrays an extraordinary feature of matter: that the elements are related to one another. We are now so familiar with the table that that feature is easily forgotten. But imagine yourself in an era before the table had been formulated.

Then you would have known of the gas oxygen and the yellow solid sulfur, and would almost certainly not have dreamt that there could be any relation between them. You would have known of the largely inert gas nitrogen and the incandescent solid phosphorus, and would not have conceived that they were related. And what about red copper, lustrous silver, and glowing gold? A family? Surely not! How, indeed, is it even possible for different forms of matter to be brothers or cousins? Even the concept of family relationships between different substances was hardly conceivable.

The Periodic Table, though, reveals that the elements are indeed related to one another. Oxygen and sulfur are cousins and stand next to each other in the table; so are nitrogen and phosphorus; copper, silver, and gold are members of the same family and lie together. Their very different appearances are superficial differences, for when the reactions they take part in and the molecules they form are investigated, it turns out that there are deep similarities between these relatives. Those similarities stem from the structures of their atoms, and to understand them it is to these atoms that we must now turn.

The structure of atoms

It was perhaps a little disconcerting for me to mention in Chapter 1 that to understand the structures of atoms it would be necessary to turn to quantum mechanics and all its incomprehensibilities. However, I did also mention that I would distil from that extraordinary theory only concepts and information that concern us. With that restriction in mind, it turns out that atoms have a rather simple structure and that it is quite easy to understand relationships between the elements and to understand, as this account unfolds, why some combinations of atoms are allowed and others not.

The basic structure of an atom consists of a nucleus surrounded by a cloud of electrons. This is the 'nuclear atom', the model of an atom first identified by Ernest Rutherford (1871–1937) in 1911. The nucleus is positively charged, the electrons are negatively charged, and it is the attraction between these opposite charges that is responsible for the existence and survival of the atom. As is well known, atoms are very small: there are over a million carbon atoms in (the printed version of) the full stop at the end of this sentence. A nucleus is even smaller: if an atom were enlarged to the size of a football stadium, the nucleus would be the size of a fly at its centre.

I shall start at the centre of the atom and work out. A nucleus consists of two types of subatomic particle: protons and neutrons. As suggested by the p and n in their names, protons are positively charged and neutrons are electrically neutral. Apart from that, they are very similar, with almost the same mass. They are tightly gripped together in the nucleus, and it requires a major effort—something like a nuclear explosion—to shake them loose. In most of chemistry, with its relatively puny releases of energy, the nucleus remains unchanged and is a passive but important participant in the processes going on around it in test-tubes, beakers, and flasks.

The number of protons in the nucleus determines the chemical identity of the atom. Thus, an atom of hydrogen has one proton, an atom of helium has two, an atom of carbon has six, nitrogen seven, oxygen eight, and so on up to livermorium, with 116. The number of protons in the nucleus is called the *atomic number* of the element. At once, we arrive at the first extraordinary feature of elements: they can be put in order according to their atomic number. No longer are elements a random jumble. They lie in a definite sequence: hydrogen, helium,…livermorium. Moreover, because the atomic number can be used as a kind of roll-call, chemists and physicists know that they have identified the elements for every atomic number up to 116 apart from (in 2014) 113 and 115. They know that none is missing except those two and whatever lies beyond 116.

The neutrons are just passengers in this roll-call. A nucleus has about the same number of neutrons as protons in the nucleus, and that number can vary slightly. As the number of neutrons does not affect the atomic number, the same element can have atoms of slightly different numbers of neutrons and therefore different masses. These different versions of the atoms of the same element are called *isotopes* because they live in the same place (*isos* = 'equal' + *topos* = 'place') in the Periodic Table. Thus, hydrogen has three isotopes: hydrogen itself (one proton, no neutrons), deuterium (one proton, one neutron), and tritium (one proton, two neutrons). The first of these isotopes of hydrogen is by far the most abundant; the nucleus of tritium barely holds together and is 'radioactive', emitting radiation as it falls apart after a few years (its 'half-life' is 12.3 years). Deuterium is 'heavy hydrogen', with each atom weighing about twice that of ordinary hydrogen. In combination with oxygen, it forms 'heavy water', which, because deuterium atoms are heavier than hydrogen atoms, is about 10 per cent heavier than ordinary water.

The atomic number, the number of protons and hence the positive charge of the nucleus, determines the number of electrons that surround it. An *electron* has the same magnitude of electric charge as a proton, but opposite in sign. Therefore, for an atom to be electrically neutral the number of electrons outside the nucleus must be the same as the number of protons inside the nucleus. That is, the number of electrons is equal to the atomic number. Thus, hydrogen (atomic number 1) has one electron, carbon (atomic number 6) has six electrons, and so on, up to livermorium with its 116 electrons. Electrons are much lighter than protons and neutrons (by a factor of nearly 2,000), so their presence barely affects the mass of an atom. They have a profound effect on the chemical and physical properties of the element, and almost all chemistry can be traced to their behaviour.

Chemists have little interest in nuclei except for their role in determining how many electrons surround them. There is one exception, the very special individual case of the nucleus of a hydrogen atom, a single proton. I shall explain its special role in Chapter 4.

As I have mentioned, all chemical reactions leave nuclei intact. In other words, chemical reactions do not change the identities of elements. At a blow, we can see why the alchemists' desperate search for means of converting lead (element 82, with 82 protons in its nucleus) into gold (element 79, with 79 protons in its nucleus) was doomed to failure: heating, stirring, banging, and stamping in frustration could not extract the tightly bound three protons from the nucleus that was necessary for 'transmutation', the conversion of one element into another. Transmutation can occur, but that, the result of *nuclear* reactions, is the domain of nuclear energy and nuclear physics. Chemists have a vital role to play in dealing with the consequences of nuclear processes, especially in preparing nuclear fuel and dealing with nuclear waste, but *chemical* reactions leave all nuclei intact, and at this stage only chemical reactions are our concern.

Electrons in atoms

My focus now turns to the hugely important properties of the electron clouds that surround a nucleus. I need to make more precise the nature and structure of those clouds, for they are not just regions of swirling mist.

Electrons surround the nucleus in layers, rather like real clouds lying above each other, but encircling the entire atom. The concept of an electron being a 'cloud' needs a quick word of explanation. The cloud is really a cloud of probability: where it is dense, the electron is likely to be found; where it is sparse the electron is unlikely to be found.

The laws of quantum mechanics ordain that up to two electrons surround the nucleus in the lowest layer, up to a further eight in the next surrounding layer, and then a further 18 in the next layer. We don't need to go beyond that, but a similar pattern with variations continues indefinitely as the number of electrons grows. This pattern means that in hydrogen a single electron surrounds the nucleus. In carbon, with its six electrons, two electrons form the lowest-level cloud and four more form surrounding clouds in the outer layer. You could think of atoms as nuclei surrounded by onion-like layers of clouds, each inner layer being completed before the next layer begins. Why there are these characteristic numbers (2, 8, 18...) for successive layers need not concern us, but is fully understood in terms of quantum mechanics.

You are now face-to-face with the explanation of the structure of the Periodic Table and the familial relationships of the elements. Keep an eye on the table at the end of this book and start at hydrogen with its single electron. Go to helium, with its two electrons. Now the first cloud layer is full and, simultaneously, we find ourselves on the far right of the table. The next electron, needed by lithium, has to become a cloud in the next surrounding layer. Stepping across the table as electrons are added, passing carbon, nitrogen, and oxygen on the way, we complete the layer at neon, another gas, like helium. The next added electron must start the next cloud layer, and brings us to sodium, on the far left of the table, an element strongly resembling lithium in the row above it, both with a single electron outside completed cores of clouds.

Everything should now be clear: the layout of the table represents the filling of the cloud layers, with one electron present in the layer on the left of the table and the layer completed on the right. For technical reasons that are fully understood, but which would be a distraction here, the order in which cloud layers are completed gets a little muddled after the first two rows of the table, and although the lengths of the rows are the numbers we have already seen, namely 2, 8, 18..., and can be discerned, they lie in a funny but understood order (the pattern of the Periodic Table is 2, 8, 8, 18, 18...).

The crucial point is that elements that lie beneath each other have very similar patterns of cloud coverage. That is the origin of family relationships: oxygen and its cousin sulfur in the row below have the same pattern of clouds, it is just that sulfur's final six electrons lie in a higher level than oxygen's final six. Likewise phosphorus's final five electrons lie at a higher-level layer than nitrogen's final five in the row above it.

It is often said that atoms are mostly empty space. That simply isn't true. The cloudlike distributions of electrons fill the whole of space around the tiny fly-in-a-stadium-sized nucleus. Admittedly the cloud is very thin in parts; but it is there and all-pervasive. The assertion that an atom is mostly empty space springs from the outmoded view that electrons are like tiny pointlike planets whizzing round the nucleus at great distances from it, with lots of emptiness in between. Quantum mechanics replaces that figure with the cloudlike distributions that I have described, clouds that, although greatly attenuated in parts, fill all space.

How atoms form bonds

The principal concern of chemistry is not so much with individual atoms but the compounds that they form by entering into a variety of liaisons with one another. There are literally millions of such liaisons that have been identified and many more that we know exist but have not been identified and named. The richness of our environment is due to this huge collection of compounds, and chemists spend most of their hours building new combinations of atoms or tearing compounds down to see how they are built. To do this effectively, they need to understand how atoms link together and what controls the links, the *chemical bonds*, that they can form to one another.

What holds atoms together to form identifiable compounds, such as water, salt, methane, and DNA? Can there be any combination of atoms, or are there reasons for Nature's restraint which chemists despite all their meddling cannot circumvent? Why is there variety in the world of substances, but apparently not random variety? These questions can be inverted: why don't all the atoms of the universe just clump together in one huge solid mass?

The answer to all these questions lies in those layers of clouds. Broadly speaking, there are energy advantages in an atom acquiring a complete cloud layer. It can do that in a variety of ways. One is to shed electrons from the outermost layer. This it is likely to do if there are not many electrons in that layer to begin with, which means that it is more likely to happen with atoms of elements on the left of the Periodic Table, at the beginning of each new row and each new layer of cloud. Alternatively, if it already has a lot of electrons in its outermost layer, then it might gain electrons from somewhere and so complete its layer. That is likely to happen if the layer is almost full, which is the case for atoms of elements on the right of the Periodic Table towards the right-hand end of a row. There is another way to complete their layers: atoms could share electrons from each other's outermost layer. That might happen when one atom is reluctant to release an electron fully because there is no energy advantage in it. That subtle mediocrity carbon forms most of its extraordinary liaisons this way.

As we have seen, atoms are electrically neutral, with the total negative charge of all its electrons matching and cancelling the total positive charge of all the protons in the nucleus. When an atom gains or loses an electron, the balance of charges is upset and the atom becomes an *ion*. An ion is simply an electrically charged atom; it is so called because it will move in response to an electric field, and 'ion' is the Greek word for going. An atom that has gained one or more electrons is negatively charged and is called an *anion*. One that has lost one or more electrons is positively charged and is called a *cation* (pronounced 'cat ion'). The 'an' and 'cat' prefixes are from the Greek words for 'up' and 'down' and reflect the fact that oppositely charged ions move in opposite directions in the presence of an electric field. I can summarize the remarks in the preceding paragraph by saying that elements on the left of the Periodic Table are likely to lose their few outermost electrons and so become cations; those on the right of the table with nearly completed outermost clouds are likely to gain electrons and so become anions.

We have come across one of the great bonding mechanisms: because opposite charges attract one another, and cations and anions are oppositely charged, it follows that atoms that form these ions will clump together into a compound. Common salt, sodium chloride, is an excellent example of this type of compound formation. Sodium (Na, from its Latin name *natrium*) lies on the left of the table, and readily releases its single outermost electron to form a sodium cation, denoted Na^+. Chlorine (Cl) lies on the right of the table, and happily accommodates an additional electron to complete its outer layer and thereby become a chloride anion, Cl^-. (Note the tiny change of name from chlorine to chloride). The ions clump together, and form sodium chloride, a solid rigid mass of ions held together by their mutual attraction. I have already emphasized that atoms are very small, and that even tiny samples of a substance contain a lot of them. You are an Atlas among stars when it comes to ions, for when you pick up a grain of salt, you are holding more ions than there are stars in the visible universe.

You are now in a position to see why salt mined in one place or extracted from a sea somewhere has the same composition as another sample mined or obtained on the other side of the world. A sodium atom has one electron in its outer layer; a chlorine atom has a single vacancy in its; so the only combination possible is for one sodium atom to bond with one chlorine atom by this process of giving up and acquiring electrons to become ions. Universally, common salt is NaCl with sodium and chloride ions present in the ratio 1:1. Compounds like Na_2Cl (ions present in the ratio 2:1) or

Na_2Cl_3 (ions present in the ratio 2:3) and so on simply can't exist. It should be becoming apparent that Nature has rules about which liaisons can form and which cannot.

The type of bonding that I have described so far is called *ionic bonding*. It typically results in rigid, brittle solids that melt only at high temperatures. The granite and limestone of our landscapes are examples of materials composed of atoms held together by ionic bonds. That we do not sink through either when we stand on them can be traced to the fact that the layers of electrons round the nuclei of their atoms, now present as ions, are complete, and the clouds of our atoms cannot occupy the same space as the clouds of their atoms. Our bones are also largely ionic, and provide a reasonably rigid framework for our organs.

Our squishy organs, our flesh, the coating of our flesh in fabrics, the fabric-analogue coating of limestone by vegetation, the upholstering of our landscapes, are all clearly of a different character. Although ions might be present, they are not responsible for the major character of these structures. Here we are in the realm where atoms are held together by completing their cloud layers by *sharing* electrons. This type of bonding is called *covalent bonding*, the 'co' indicating cooperation and the 'valent' derived from the Latin word for strength: *Valete*! was the Roman 'Goodbye! Be strong!'.

A simple example of covalent bonding is that responsible for the structure of a water molecule, which just about everyone knows is H_2O. Oxygen, with its six outermost electrons can accommodate two more electrons to complete its outermost cloud layer (which can hold, remember, a maximum of eight electrons). A hydrogen atom can provide one electron, and can complete its own outermost cloud layer (the only one it has) by acquiring one more electron (that first, innermost layer, remember, can accommodate only two electrons). Sharing can be complete provided two hydrogen atoms are content to share two electrons with oxygen: the hydrogen atoms each get a share in two electrons and the oxygen atom gets a share in eight electrons. At once, we see that water cannot be H_3O or HO_2: H_2O is the only bonding pattern that results in complete outermost cloud layers for all the atoms. Ammonia, NH_3 (where N denotes nitrogen) also falls into place, because a nitrogen atom has five electrons in its outer layer, and so needs three more to complete its layer. That is satisfied by the presence of three hydrogen atoms willing to share an electron each. Methane, CH_4, falls into place too, because carbon has four vacancies.

You, like chemists, need to be aware of one very important distinction between ionic and covalent bonding. Ionic bonding results in huge aggregates of ions: essentially chunks of substance. Covalent bonding commonly results in discrete atomic assemblies, like H_2O. That is, covalent bonding results in individual molecules. This distinction is hugely important, and you need to keep it in mind. It is for this reason that all gases are molecular, such as oxygen (O_2 molecules) and carbon dioxide (CO_2 molecules); there is no such thing as an ionically bonded gas! Even if such a gas were formed, all the ions would immediately clump together as a solid. Just about all substances that are liquid at normal temperatures are molecular, as the molecules need to be able to move past one another and not be trapped in place by a strong attraction to their neighbours. Water is an obvious example; gasoline another.

Covalent bonding can result in solids, so you should not infer that every solid is ionic: all ionic compounds are solids at ordinary temperatures but not all solids are ionic. An example of a covalently bonded solid compound is sucrose, a covalent compound of carbon, oxygen, and hydrogen with the composition $C_{12}H_{22}O_{11}$ with the atoms in each molecule linked together by covalent bonds into an intricate web.

One very important aspect of covalent bond formation is the overriding importance of pairs of electrons. One of the greatest chemists of the 20th century, Gilbert Lewis (1875–1946), identified its importance, but it remained for quantum mechanics to provide an explanation. As far as we are concerned, each shared pair of electrons counts as one covalent bond, so it is easy to count the number of bonds that any atom has formed simply by counting the number of pairs of electrons that they share. One shared pair counts as a 'single bond' (denoted –), two shared pairs between the same two atoms counts as a 'double bond' (denoted =), and three shared pairs counts as a 'triple bond' (denoted ≡). Only very rarely does sharing proceed any further, and so these three types of sharing are all we need to know about. Each hydrogen atom in H_2O is joined to the oxygen atom by a single bond. Carbon dioxide is a molecule with two sets of double bonds, and can be denoted O=C=O. Triple bonds are much rarer, and I shall not discuss them further except to mention that the gas acetylene of oxyacetylene welding, H–C≡C–H, is an example.

The question lurking behind this account is why two electrons (an 'electron pair') are so fundamental to covalent bond formation. The explanation lies deep in quantum mechanics. A hint of the reason is that all electrons spin on their axis. If two electrons lock their spins together by rotating in opposite directions, then they can achieve a lower energy. Another manifestation of the importance of this spin-locking is the fact that, as we have seen, the cloud layers each hold an even number of electrons (2, 8, etc.). The French words for an unpaired electron, an *electron célibataire*, is a perhaps typically Gallic allusion to the importance of pairing.

Metals

I have concealed from view so far the existence of a third type of bond. The majority of elements are metals: think iron, aluminium, copper, silver, and gold, and metals play a very special role in chemistry, as we shall see. A block of metal consists of a slab of atoms, but are those atoms held together by ionic or covalent bonds? We are immediately confronted by a problem. All the atoms in the block are the same, so it is unlikely that half will form cations and the other half anions, so ionic bonding is ruled out. If all the atoms were bonded covalently, we would expect a rigid solid (like diamond, in which the carbon atoms are in fact so bound); but metals can be beaten into different shapes (they are 'malleable') and drawn out into wires (they are 'ductile'). They are also lustrous (reflective of light) and conduct electric currents, a stream of electrons.

Metal atoms are bound together by *metallic bonding*. That is not just a tautology. The clue to its nature is the fact that all the metals lie towards the left-hand side of the Periodic Table where, as we have seen, the atoms of the elements have only a few electrons in their outermost cloud layers and which are readily lost. To envisage metallic bonding, think of all these outermost electrons as slipping off the parent atom and congregating in a sea that pervades the whole slab of atoms. The cations that are left behind lie in this sea and interact favourably with it. As a result, all the cations are bound together in a solid mass. That mass is malleable because, like an actual sea, it can respond readily to a shift in the positions of the cations in the mass when they are struck by a hammer. The electrons also allow the metal to be drawn out into a wire, by responding immediately to the relocation of the cations. As the electrons in the sea are not pinned down to particular atoms, they are mobile and can migrate through the solid in response to an electric field. Metals are lustrous because the electrons of the sea can respond to the shaking caused by the electric field of an incident ray of light, and that oscillation of the sea in turn generates light that we perceive as reflection. When we gaze into the metal coating of a mirror, we are watching the waves in the metal's electron sea.

The chemistry lesson at this stage in our account is that the elements that are metals in their natural state are the ones that can readily lose electrons from their outer layers. These elements are therefore also the elements that form cations when anion-formers are present and able to accept the discarded electrons. Elements on the far right of the Periodic Table are electron acquirers, as they have one or two gaps in their outer cloud layers and can accommodate incoming electrons, those donated by the atoms that form cations. Ionic compounds (bear in mind sodium chloride) are therefore typically formed between a metallic element on the left of the table with a non-metallic element on the right of the table.

With that summary in mind, you are starting to think like a chemist, being able to anticipate the type of compound that a combination of elements is likely to be, and beginning to anticipate its properties. You are also beginning to understand how the Periodic Table relates to the properties of the elements and the compounds they form, and how the family relationships between neighbours, which spring from the cloudlike electrons and the periodic repetition of analogous arrangements, are displayed in practice.

Where we are, and the next step

Such are the central principles of chemistry as far as structures are concerned. They boil down to the existence of atoms, an acknowledgement of their structures, and the behaviour of electrons. Our next concern is with the 'carrot and the cart' of chemistry: energy.

Chapter 3
Its principles: energy and entropy

Atoms are one great river of understanding in chemistry; the other river consists of energy. To understand why and how reactions take place and why and how bonds form in all their variety, chemists think about the energy changes that take place when processes occur. Chemists are also interested in energy for its own sake, as when a fuel is burned or food, a biological fuel, is deployed in an organism. As I remarked in Chapter 1, the study of energy and the changes that it can undergo is the world of thermodynamics, to which we now turn.

I have written extensively on the laws of thermodynamics and do not intend to reprise my discussion here. As I did for quantum mechanics in Chapter 2, I shall distil the essence of what is necessary and which chemists typically keep in mind or at least the back of their minds while going about their business.

Some thermodynamics

The essence of chemical thermodynamics is that there are two aspects of energy that it is necessary to keep in mind: its quantity and its quality. The *First Law* of thermodynamics asserts that the total energy of the universe is constant and cannot be changed. The energy can be parcelled out in different ways and converted from one form to another, but no process can change its total quantity. Thus, the First Law sets the legal boundaries for change: no change can occur that would alter the total amount of energy in the universe. The *Second Law* of thermodynamics asserts that the quality of energy degrades in any natural change. This law is expressed more formally in terms of the *entropy*, a measure of the quality of energy in the sense that the higher the entropy the lower is its quality, and stated as 'the entropy of the universe tends to increase'. In a refined meaning of 'disorder', entropy is a measure of disorder, with greater disorder implying greater entropy. The Second Law can be regarded as a summary of the driving power of natural change, including chemical reactions, for only reactions that result in the degradation of the quality of the total energy of the universe can occur naturally. In short, with increasing disorder in mind, things get worse. A summary of thermodynamics, the core of its essence, is therefore that the First Law identifies the feasible changes from among all possible changes (no change in total energy) and the Second Law identifies the natural changes from among those feasible changes (the entropy must increase).

The role of energy

Chemists deploy these two concepts in a variety of ways. In their conventional thinking they adopt the view that bonds form or are replaced by new bonds in the course of a reaction if that reorganization of atoms results in a reduction in energy. That remark, though, to a fusspot like me, is quite wrong, but like many false statements it is a handy and memorable rule-of-thumb. It is wrong because the legal authority of the First Law rules against it: the total energy cannot change. The correct explanation is that if a process, such as bond formation, releases energy into the surroundings, then that represents a degradation of energy as it spreads and becomes less readily available: the release increases the entropy of the universe and so is a natural process. That the rule-of-thumb 'it lowers the energy' works most of the time is due to the fact that the spread of energy so released results in an increase in entropy. Working chemists quite sensibly use the rule-of-thumb all the time and I shall follow them. However, I shall keep my fingers crossed when I use it, and inwardly say to myself, a little like Galileo's whispered apocryphal *eppur si muove* ('and yet it moves') concerning the motion of the Earth around the Sun, that it is really entropy going up rather than energy going down.

A bond between atoms forms if (fingers crossed) it results in a reduction in energy. The type of bond that forms, ionic (attraction between ions) or covalent (shared electron pairs), depends on whether more energy is released by the total transfer of an electron from one atom to the other to result in ions, or by partial release and sharing. Thus, whether two elements form an ionic or a covalent compound can be assessed by considering the energy changes that accompany the various types of bond formation.

The same is true of the characteristic *valence* of an element, the typical number of covalent bonds that it can form. Valence is another aspect of its chemical personality and family relationship to its neighbours and is implied by its location in the Periodic Table. We saw in Chapter 2 that oxygen, with its two gaps in its outermost cloud layer, can complete that layer by reaching agreement with two hydrogen atoms to form H_2O, specifically H–O–H, indicating a valence of 2. Any further attachment of hydrogen atoms would require electrons to occupy a new outer cloud layer far from the nucleus, and there would be no energy advantage in doing so. Forming fewer bonds would not reap the advantage of forming two. Therefore, on energetic grounds, the valence of oxygen is expected to be 2. Also in Chapter 2 we saw another example of that valence in oxygen's combination with carbon in carbon dioxide, CO_2, specifically O=C=O, where it also displays a valence of 2. As can be seen in this case too, carbon displays its typical valence of 4, just as it does in methane, CH_4.

Now we can see how the location of an element in the Periodic Table indicates its characteristic valence: carbon's typical valence is 4, its neighbour nitrogen is 3, and nitrogen's neighbour oxygen is 2. Much the same can be said of their neighbours in the row below: silicon's valence is typically 4, phosphorus's is 3, and sulfur's is 2. Once again, we are seeing how energy considerations in collaboration with concepts of atomic structure—particularly the completion of their cloud like layer structure—accounts for similarities between neighbouring elements.

Keeping track of energy

Energy is released in many chemical reactions, as in the combustion of natural gas or gasoline. The process is not simply the release of energy when bonds form, because the starting materials, such as methane, already have atoms bonded together. In many reactions, and here I shall focus on combustion, bonds must be broken and new bonds formed. The energy released is the difference of the two contributions. For instance, in the combustion of methane, due to its reaction with oxygen, O_2, the four carbon–hydrogen bonds of methane and the bonds linking the two oxygen atoms in oxygen must all be ripped apart, which takes a lot of energy, before new carbon–oxygen bonds in carbon dioxide and hydrogen–oxygen bonds in water are formed, which releases energy. Only if the energy released in the subsequent bond formation exceeds the energy required for initial bond-breaking will the combustion release energy as heat. If the balance were the other way round, burning methane would result in refrigeration!

Chemists use thermodynamics to keep track of these individual changes in energy, and to assess the net change that takes place in a reaction. For this purpose, they use an assessment of the quantity of energy available from a reaction as heat that is called the *enthalpy*. The name comes, evocatively, from the Greek words for 'heat inside'. There are good technical grounds for distinguishing enthalpy from energy, but for our purposes we can think of enthalpy as just another name for the energy trapped in compounds and available as heat.

In a so-called *exothermic reaction*, energy is released as heat and the store of enthalpy decreases. All combustions are exothermic, and in the combustion of methane the enthalpy of methane + oxygen falls to the enthalpy of carbon dioxide + water, the difference escaping as heat. Chemists assess the efficiency of fuels by considering the enthalpy changes that accompany their combustion, with full reservoirs of enthalpy being preferred as more heat is available from a given amount of fuel. The study of enthalpy and the release of heat in chemical reactions is called *thermochemistry*. It makes a substantial contribution to our understanding of foods and fuels and is also used to gather data for more general thermodynamic discussions.

Most reactions, not only combustions, are exothermic, with the starting materials collapsing into the lower-enthalpy products of the reaction and thereby achieving lower enthalpy overall. It is perhaps easy to understand that many reactions proceed in the direction of lower enthalpy, just as the finger-crossing rule-of-thumb about energy suggests. However, here is a puzzle, a puzzle that left 19th century chemists totally nonplussed: some reactions move upwards in enthalpy naturally. Reactions that absorb heat and increase their store of enthalpy are called *endothermic reactions*. There are not many common ones that occur naturally, but the fact that there is even one raised the collective puzzled eyebrows of the 19th century chemists, for how, they wondered, can anything run naturally uphill, in this case, uphill in enthalpy?

They didn't know about entropy, and they took literally the rule-of-thumb about things falling naturally to lower energy. Chemists now know that entropy determines the direction of reactions, and *provided the entropy increases*, the reaction can either travel uphill or downhill in enthalpy. To understand why, we have to remember that entropy is a measure of the quality of energy.

When energy spreads into the surroundings of a reaction flask and becomes dispersed, the entropy goes up, so it should be easy to understand why exothermic reactions are so common. However, we need to think about what is going on inside the flask. Suppose that in the course of a reaction energy flows into the flask: the entropy goes down because now the energy is localized, less dispersed, and more readily available: it has become of higher quality. Suppose, though, that at the same time a great deal of disorder is generated within the flask. Now the total entropy of the universe might increase despite the energy becoming more localized. If that happens, then the endothermic reaction will occur naturally.

Where the 19th century chemists went wrong was to suppose that, like Newton's apple, reactions rolled down in enthalpy; what 21st century chemists know is that reactions roll up in entropy: disorder increases; things get worse. Often the two lead to the same conclusion, but in all cases entropy is the property to consider. Increasing entropy is the signpost of change, and sometimes it points in an endothermic direction. If you still continue to want to think, from your familiarity with gravity, that the natural direction of change is 'down', then think that natural change it is invariably down in the quality of energy.

The rates of reactions

We now know where a reaction goes: the signpost of natural change is towards higher entropy of the universe, the degradation of energy. There are two associated questions. One is how fast it goes to wherever it is going, and the second is what route it takes to get there. I shall deal with the first question here and tackle the second in Chapter 4.

Chemists take a great deal of interest in the rates of chemical reactions as there is little point in knowing that they can, in principle, generate a substance in a reaction but that it would take them millennia to make a milligram. The study of reaction rates is called *chemical kinetics*. We shall see that energy is a crucial component of the explanation of the wide range of rates that are observed. That range is indeed very wide: some reactions are complete in fractions of seconds (think explosions); others take years (think corrosion).

Chemists measure reaction rates in a straightforward way, by monitoring the change in amount of a product over time. They do these measurements for a variety of reasons. One, the most basic, is simply to know what concentration to expect at any given moment. More significantly, especially for industrial applications, they may wish to find the conditions that result in products being formed at the optimum rate. A third reason is to discover what is called the *mechanism* of the reaction, the sequence of changes at an atomic level that converts the starting materials, the 'reactants', into the final product. Very detailed information of the last kind is obtained by firing one stream of molecules at another and monitoring the outcome of the collisions that take place.

My concern here is with the role of energy in determining the rate of a reaction. We have seen that there may be a natural tendency for a reaction to occur, so the question arises why all reactions aren't over in a flash. This question is supremely important, for the slow, restrained development of products in many cases allows for the subtle operations that constitute life: if biological reactions were all over in a flash we would all instantly be goo.

Chemists have identified the existence of a barrier to instant reaction. By making measurements on the effect of temperature on the rates of reactions, they have identified the need for molecules to acquire at least a minimum energy, called the *activation energy*, before the atoms of the reactants are able to rearrange into products. This requirement is easiest to understand for reactions in gases, where molecules are ceaselessly undergoing collisions with one another with various energies of impact. Only highly energetic impacts between really fast molecules bring enough energy to loosen the bonds holding atoms in their initial arrangements and enabling them to settle into new ones. As the temperature is raised, molecules move faster and a higher proportion of the collisions occur with at least this minimum energy, so the rate of the reaction increases. Some activation barriers are very high, and hardly any collisions are sufficiently energetic to result in reaction at normal temperatures. The reaction of hydrogen and oxygen is an example: the two gases can be stored together indefinitely at normal temperatures but explode at high temperatures or when a spark provides sufficient energy locally to set the reaction in train.

Much the same requirement of a minimum energy applies to reactions in solution too, including those in the watery interiors of living things. In this environment, molecules do not hurtle through space and collide: they jostle through the fluid, meet, and might jostle away unchanged. However, there is a chance that when two reactant molecules are together, they are jostled so violently by the surrounding water molecules that their atoms are eased apart and can rearrange into products. The chance that sufficiently violent jostling will occur increases sharply with increasing temperature, so even reactions in fluid environments go faster when they are heated. Fireflies, for instance, flash more rapidly on warm nights than on cool nights; we heat to induce the reactions in the kitchen that we call 'cooking' foods.

In many instances a reaction can be made to go faster by introducing a *catalyst*, a substance that increases the rate of the reaction but is otherwise unchanged. The Chinese characters for catalyst form the word 'marriage broker', which captures the sense of its role very well. A catalyst acts by providing a different pathway—a different sequence of atom migrations and bond formations—for a reaction, a pathway with a lower activation barrier. Because the activation energy is lower, more successful encounters between reactants take place at ordinary temperatures and the reaction is faster. Catalysts are the lifeblood of the chemical industry, where the efficient, rapid production of desired substances is essential and the success of an entire industry depends on the identification of the appropriate catalyst. A point to note is that there is no such thing as a 'universal catalyst', and each reaction must be studied individually and an appropriate catalyst devised. Another point is that not all reactions can be catalysed: in many cases we have to live with Nature's decision about the rate.

Catalysts are essential for the functioning of our bodies. *Enzymes* (a word derived from the Greek word *zyme*, to leaven) are protein molecules that function as catalysts and control with considerable specificity and effectiveness just about all the chemical reactions going on inside us. Life is the embodiment of catalysis.

The nature of equilibrium

A very important aspect of reaction rates concerns what is going on when a reaction has completed and change is no longer apparent. Chemists say that the reaction has reached *equilibrium*. In many cases, barely a single molecule of the starting material remains, but in many cases the reaction seems to stop before the starting materials have all been used. An example of the latter is the hugely economically significant reaction between nitrogen and hydrogen for the synthesis of ammonia (NH_3) in the 'Haber–Bosch process', which lies at the head of processes that include the manufacture of much of the world's agricultural fertilizer. That reaction seems to come to a stubborn stop with only a small fraction of the nitrogen and hydrogen converted into ammonia, and however long we wait, and however much catalyst we shovel in, no further change occurs. The reaction has reached equilibrium.

Equilibrium is only an apparent cessation of reaction. If we could monitor an equilibrium mixture at an atomic level, we would find that it is still a turmoil of chemical activity. Products are still being formed when a reaction is at equilibrium, but they are decaying back into the starting material at a matching rate. That is, chemical equilibrium is a *dynamic* equilibrium, in which forward and reverse processes are occurring at matching rates so that there is no net change. In the synthesis of ammonia, ammonia molecules are still being formed at equilibrium, but are being ripped apart into nitrogen and hydrogen at the same rate as they are being formed and there is no net change.

The important consequence of chemical equilibrium being dynamic and not just dead is that it remains responsive to changes in the conditions. Thus, even though certain reactions inside our bodies might have reached equilibrium, they are responsive to changes in temperature and other factors, and it is that responsiveness that keeps us alive. 'Homeostasis', the delicate and complex balance that keeps bodies alive and alert, is a manifestation of this dynamic, responsive, chemical equilibrium. As to the industrially all-important synthesis of ammonia, the fact that the equilibrium is dynamic rather than dead gives chemists and industry hope that perhaps the equilibrium can be manipulated and the yield of ammonia improved. That was the prospect confronting the chemist Fritz Haber (1868–1934) and the chemical engineer Carl Bosch (1874–1940) back at the beginning of the 20th century, who in due course discovered that with an adroit choice of catalyst and by working at high pressures and temperatures, they could bend the equilibrium to their will. In so doing, they fed the world.

Where we are, and where we are going

We have now seen that energy is both the carrot and the cart of chemical reactions, and so can finally unwrap the meaning of my delphic remark at the end of Chapter 2. Energy, its dispersal in disorder, is the carrot: the driving power of chemical reactions. Energy, the need to overcome the barriers between reactants and products, is also the cart, in the sense of holding back free unrestrained flight towards the carrot.

I have said hardly anything about how reactants actually undergo the atomic rearrangement that leads to products. Unravelling and understanding those changes, and making use of them to bring about amazing and almost magical transformations, lies at the heart of practical chemistry, and is the next step in our journey.

Chapter 4
Its reactions

Whenever anyone thinks of chemistry, they think of its reactions, reactions that flash, bang, change colour, or stink. They are aware that reactions go on in chemical plants, that the combustion of a fuel and the manufacture of plastic, paint, or pharmaceuticals are reactions. Perhaps some correctly think of cooking as causing reactions, and most are probably at least vaguely aware that we ourselves are elaborate test-tubes who are alive as a result of the myriad reactions within us. But exactly what are reactions? What is going on when chemists stir and boil their liquid mixtures, pour one liquid into another, and generally go about their seemingly arcane activities in laboratories?

They are coaxing atoms to exchange partners. The starting stuff, the 'reactants', consists of atoms in one state of combination; the stuff that is produced, the 'products', consists of the same atoms but in a different state of combination. The shaking, stirring, and boiling is bringing about that change from one state of combination to another, prising atoms apart in one kind of molecule and encouraging them to form different kinds of molecules. In some cases, the atoms of the reactants immediately tumble into the desired new arrangement, whereas in others, the chemist must scheme and seduce, devising elaborate coaxings through a sequence of subtle steps. Combustion and explosion might stem from a spark; to generate an intricate weblike pharmaceutical molecule might take thought, luck, time, and careful, sophisticated, erudite planning.

A chemical laboratory is full of specialized equipment, a lot of which is there to determine whether the product of a reaction is what the chemist hopes or thinks it is. I explain some of its functions in Chapter 5. A lot of it is directly involved in the business of atom-coaxing and separating the chemical wheat from the chaff, the desired product from the waste. There are test-tubes, flasks, beakers, distillation apparatus, filtering apparatus, and various heaters, shakers, and stirrers. Despite this bewildering (and expensive) array of apparatus, through a chemist's eyes there are only a small number of processes going on at an atomic level: four, to be precise. It is in fact worth pausing at that remark, and to realize that all the wonders of the world, both natural and synthetic, are spun from a handful of elements and four ways of manipulating them.

In the rest of this chapter I shall introduce you to those four fundamental types of reaction. In some cases, they conspire together and their collaboration at first sight seems to be of a new type of reaction, but when that conspiracy is picked apart, there they are.

Proton transfer: acids and bases

Chemists discovered the fundamental particle known as the 'proton' long before the physicists had pinned it down, but the chemists did not realize that they had done so. A proton, remember from Chapter 2, is the tiny, singly charged nucleus of a hydrogen atom. Its low charge (which means that it is often only loosely gripped by a neighbouring atom in a molecule) and low mass (which makes it nimble) mean that a hydrogen atom that is part of a molecule might suddenly find that its nucleus, the proton, has slipped away and become embedded in the electron clouds of a more welcoming nearby molecule. That—the transfer of a proton from one molecule to another—in a nutshell, is one of the four great fundamental types of reaction.

We are in the world of acids and alkalis. Although the early chemists were familiar with acids, it took them a long time to realize that an acid is a compound with hydrogen atoms that have little control over their nuclei and are apt to lose them. Acids at one time were recognized, as their name suggests (Latin *acidus*: sour, sharp), by their sharp taste. Chemists who survived that hazardous test (now existing in a more palatable way in our response to the tang of vinegar, soda, and cola drinks) had no idea that what was tickling their tastebuds were protons. That recognition came as late as 1923, when the British chemist Thomas Lowry (1874–1936) and the Danish chemist Johannes Brønsted (1879–1947) independently proposed that an acid is any molecule or ion that contains hydrogen atoms that can release their proton nucleus to another molecule or ion. Not all molecules that contain hydrogen can act in this way, as the proton may be too heavily embedded in the electron clouds, but various classes of molecule can, especially if other atoms in the molecule can draw the electron cloud away from the proton and enable it to escape. Acetic acid, the acid in vinegar, is one such compound; others include hydrochloric acid (HCl) and sulfuric acid (H_2SO_4). If you ever see H written first in a formula, that is an indication that it can release its proton and act as an acid. (What about H_2O?, you might be thinking: wait and see).

One hand cannot clap alone. If there is a proton donor (an acid), presumably there must be a proton acceptor, a molecule or ion to which the liberated proton can attach and burrow itself into the electron cloud. This is where alkalis come in (the name comes from the Arabic *al qaliy*, the ashes, for wood ash is a source of alkali).

The test for an alkali used to be just as hazardous as that for an acid: in this case, an alkali has a soapy feel. We now know that alkalis turn fats into soaps, so in the test the fats on the finger of the tester were being turned into soap. Needless to say, chemists have more survivable and sophisticated tests now. The underlying reason for the ability of alkalis to turn fats into soap is the presence in them of *hydroxide ions*, OH^-, which are species that can attract and keep protons, in the course of which becoming water molecules, H_2O.

Here is a tiny technical point that I really do need to introduce. Chemists now refer to a proton-accepting molecule and ion as a 'base'. Thus, OH^- is a base. They keep the term 'alkali' for bases dissolved in water. So, for instance, sodium hydroxide, NaOH, dissolves in water, separating into Na^+ ions and OH^- ions. It is therefore a source of the base OH^- and the solution is an alkali. I shall use the term 'base' from now on, because it is more general than alkali (a molecule or ion doesn't need to be present in water to be a base).

Why the name 'base'? When hydrochloric acid reacts with sodium hydroxide solution, salt (sodium chloride) and water are formed when the acid's proton skips across on to the OH^- ion provided by the sodium hydroxide. When instead sulfuric acid reacts with sodium hydroxide solution, sodium sulfate and water are formed when the acid's proton skips across on to the OH^- ion provided by the sodium hydroxide. We are building different compounds, sodium chloride and sodium sulfate, on foundations of the same base, sodium hydroxide: hence the name.

Incidentally, both sodium chloride and sodium sulfate are called *salts*, the general class of these ionic substances formed by the reaction of an acid and a base taking its name from a common exemplar, namely common salt, sodium chloride. That is a common feature in chemistry, where the name of one type of compound inspires the name of a whole related class.

A large number of reactions are reactions between acids and bases, their common feature being that a proton is transferred from the acid to the base. Among the most important are reactions going on inside organisms, including corn, oak trees, flies, frogs, and us, for many enzyme-controlled biochemical reactions, such as those involved in the metabolism of food and respiration, are of this kind. In fact, you could regard life as one long, highly elaborate titration!

One reason for the importance of proton transfer, acid–base reactions is that the presence of the arriving proton with its positive charge distorts the electron cloud of the base, perhaps exposing an atom nearby in the molecule to attack by other atoms as the electron cloud around it is pulled away. Thus, proton transfer prepares atoms and the bonds that hold them to attack and then further reaction. This preparation for attack is a major role of acid–base reactions in our bodies, with enzymes preparing smaller molecules for digestion or modification.

I suspected earlier that you might worry about water and its formula H_2O, which with its leading H atoms might suggest that it is an acid. It is. When you drink water you are drinking almost 100 per cent acid. Water is also a base. You should know that when drinking water you are drinking almost 100 per cent pure base. I need to explain this alarming revelation. Although alarming, the acceptance of the fact that water is both an acid and a base is central to the way that chemists think about it, the solutions it forms, and the reactions it undergoes.

Think of yourself as a water molecule in a glass of water, surrounded by other water molecules in a dense, jostling crowd. One of your hydrogen atom's protons can slip out of you and stick on to a neighbour. That transfer implies that, being a proton donor, you are an acid. Your neighbour, who accepted the proton, is behaving as a base. The loss of a proton leaves you as an OH^- ion, a hydroxide ion; the gain of a proton makes your neighbour an H_3O^+ ion, which is called a *hydronium ion*. The incoming proton is like a hot potato, and it is immediately passed on to one of your neighbour's neighbours. Likewise, your negative charge can pull a proton out of one of your neighbours, rendering you H_2O again. This ceaseless turmoil of passing on a proton and flickering from OH⁻ to H_2O to H_3O^+ goes on throughout the liquid. The actual concentrations of the OH⁻ and H_3O^+ ions is very, very small, so when you look at a glass of water you should think of it as overwhelmingly H_2O molecules, but with just a few OH⁻ and H_3O^+ ions throughout it, with identities that are ceaselessly changing as protons hop around. Each H_2O molecule, though, is an acid (a proton donor), and each H_2O molecule is a base; that is why I said that water is a nearly pure acid and a nearly pure base.

Electron transfer: oxidation and reduction

The electron was discovered by the physicist J. J. Thomson in 1897. Chemists had unwittingly shifted it around for decades before that, with Michael Faraday (1791–1867) its arch-shifter: clearly, even he did not know what he was doing. The transfer of an electron from one molecule to another is the second of the four great fundamental reactions, with a great deal stemming from the migration of this little fundamental particle. Electron transfer, for instance, is the basis of great industries, such as steel-making. It is also responsible for the collapse of their artefacts through corrosion.

I need to introduce you to the terms 'oxidation' and 'reduction', and then explain how electron transfer plays a role in them. Oxidation sounds as though it means what it says: namely reaction with oxygen. However, although in science a term might have started life in common usage, it often captures more of the landscape by becoming generalized. We saw that a moment ago in the generalization of the term 'salt' from a single exemplar to a whole class of related compounds. So it is with oxidation too.

Let's take a simple example. Most of us have seen the bright light emitted when a strip of magnesium (Mg) burns in air. In this reaction, the magnesium metal combines with oxygen to form magnesium oxide, an ionic solid consisting of Mg^{2+} ions and O^{2-} ions. The energy released in the reaction is emitted as light and heat. The crucial point to note, however, is that each Mg atom of the metal has lost two electrons and has become a doubly charged Mg^{2+} ion. A similar reaction, but one far less familiar, occurs when magnesium burns in chlorine gas, when the product is magnesium chloride. That compound consists of Mg^{2+} ions and Cl^- ions. As in the first reaction, the crucial change is that each Mg atom has lost two electrons to become an Mg^{2+} ion. No oxygen is involved in the second reaction, but the same process, the removal of electrons, has occurred. Chemists now regard the second reaction simply as an oxidation of a general kind, and define oxidation as *the loss of electrons*. It is sometimes quite difficult to identify electron loss, such as in the combustion of a hydrocarbon fuel, but they have ways of doing so, and whenever electron loss takes place they call it an oxidation even though oxygen itself might not be involved in any way.

We saw when discussing the reactions between acids and bases that one hand cannot clap alone: if there is a proton donor (the acid), there must be a proton acceptor (the base). One hand cannot clap alone in an electron transfer reaction either, and the electrons lost in an oxidation must end up somewhere. That is where reduction comes in.

In the old days (I am being deliberately vague), reduction referred to the extraction of a metal from its ore: the ore was *reduced* to the metal. This process occurred, for instance, in that great hulking giant icon of the industrial revolution, the blast furnace, in which iron ore (an oxide of iron) reacts with carbon and carbon monoxide to form the molten iron (Fe, from the Latin *ferrum*) that dribbled out of the base of the furnace and went on for a future as various kinds of steel. Iron oxide consists of Fe^{3+} ions and O^{2-} ions. Iron metal consists of Fe atoms. With that in mind it is easy to see what has happened in the reduction of the ore:

electrons have attached to each Fe^{3+} ion to neutralize its charge and form Fe atoms.

The attachment of electrons to an atom is now taken as the definition of reduction, even though the reaction might have nothing (except that feature) to do with the reduction of an ore to a metal. Thus, in the combustion of magnesium in oxygen, the oxygen molecules receive the electrons released in the oxidation of magnesium and become O^{2-} ions: the oxygen is reduced. In the oxidation of magnesium by chlorine, the chlorine molecules receive the released electrons and become Cl^- ions: the chlorine is reduced. Whenever electrons released are transferred to an atom, that atom is said to be reduced.

We now have both hands clapping in an electron transfer reaction: oxidation (electron loss) always occurs with reduction (electron gain). Chemists recognize the need for these two hands and commonly refer not to an oxidation reaction alone, nor to a reduction reaction alone, but to a 'redox' reaction. (They don't, so far, extend that type of compressed naming to 'basid' reactions, combined base and acid hand-clapping).

Redox reactions are hugely important. We have already seen that they stand at the head of the steel chain, when iron is won from its ores. The reverse of that winning is the process of corrosion, when the ion artefacts are lost in the redox reactions that we call corrosion: when iron is oxidized by water and the oxygen of the air and reverts to its oxide. The combustion reactions that drive our vehicles are redox reactions, in which the hydrocarbon fuel is oxidized to carbon dioxide and water by reaction with oxygen (which is itself reduced).

The reactions that take place in the batteries that power our laptops, tablets, phones, and increasingly vehicles are redox reactions. Batteries are so important for the modern world as portable sources of electric current that it is worth a moment or two to understand the general principle of their operation and how they harness redox reactions.

We have seen that in an oxidation electrons are released and that in a reduction they are acquired. In a battery, the release and acquisition are spatially separated. Electrons are released into an electrode, a metallic contact, in one region of the battery, travel through an external circuit, and then attach to the species undergoing reduction at a second electrode elsewhere in the battery. Thus, the redox reaction, the joint oxidation and reduction reactions, proceed, and in doing so, the flow of electrons from one electrode to the other is used to drive whatever electrical equipment is attached to the device. Modern batteries use a range of redox reactions to bring about this electron flow, ranging from the heavy lead–acid batteries in vehicles to the light lithium-ion batteries in laptops, tablets, and phones.

Redox reactions can also be forced to take place against their natural direction by driving electrons into a reaction mixture through an electrode. This is the process of *electrolysis*, the process of causing chemical reaction by passing an electric current. Electrolysis is the principal way of extracting aluminium (Al) from aluminium oxide. A powerful current is forced into a cell containing aluminium oxide dissolved in a special solvent, and the electrons that enter the cell are forced on to the Al^{3+} ions, forming Al atoms. Electrolysis is also used to purify copper and to deposit metals, such as chromium, on to the surfaces of other metals.

One feature that distinguishes the electron transfer of redox reactions from the proton transfer of acid–base reactions is that because electrons are intimately involved in bonding, the migration of an electron from one molecule to another can drag with it several other atoms. We have seen a little of that, without drawing attention to it, in combustion reactions, where in the course of the oxidation of a hydrocarbon molecule, carbon, oxygen, and hydrogen atoms are dragged around as the electrons migrate between the molecules, and the hydrocarbon and oxygen molecules are reassembled into carbon dioxide and water molecules. This difference is hugely important in the reactions of organic chemistry, where clever use of atom-dragging redox reactions can be used to construct intricate structures.

It is partly due to this ability of migrating electrons to drag atom baggage with them that redox reactions are so important in biology: they keep the biosphere (including that small part, you) alive and vibrant. Photosynthesis, the process by which sunlight is captured and used to power the formation of carbohydrates in green plants, is a chain of electron transfer reactions, which have the overall effect, when atom-dragging is taken into account, of using the hydrogen atoms of water and the carbon and oxygen atoms of carbon dioxide, to construct carbohydrates, including starch and cellulose. In the form of organic corrosion we call digestion, those redox-formed carbohydrates are mined for their carbon and hydrogen atoms in a sequence of redox reactions we call respiration and metabolism.

Radical reactions

The third kind of reaction takes place when radicals meet. You need to know that a *radical* (or 'free radical') is a molecule with an odd number of electrons. We have seen that electrons pair together when bonds form, so a radical is a molecule in which all the electrons except one have paired and hold the atoms together, and there remains one unpaired electron. A radical is commonly denoted R· or ·R, the dot representing the unpaired electron.

Most radicals are aggressively reactive and do not survive for long in the wild. In some cases, two radicals might collide, and clump together as their unpaired electrons pair and bind the two radicals together to form a conventional molecule with an even number of electrons: R· + ·R → R–R. This kind of process occurs in flames, which are environments rich in radicals because the stress of high temperatures rips molecules apart, sundering the electron-pair bonds. Indeed, one form of fire-retardant is a substance that gives rise to radicals when heated. These radicals, lets denote them ·X, latch on to the radicals that are propagating the flame and quench their chemical aggressiveness, R· + ·X → R–X, so that the flame peters out.

Other radicals are of great commercial importance, for they are involved in the formation of many plastics. The general idea behind this process, which is called *polymerization*, is that when a radical R· attacks an ordinary molecule M it might attach to it. However, the outcome is again a molecule, now RM·, with an odd number of electrons, so it is also a radical. That radical can go on to attack another M molecule and attach to it. The result is still a radical, now RMM·. In other words, there can be a *chain reaction*, a chain of processes that propagates indefinitely, to give a long snaking RMM...M· radical, or until two such radicals collide, stick together by electron pairing, and so terminate the chain.

Those ubiquitous plastics polythene (polyethylene), polystyrene, and PVC (polyvinyl chloride) are made in this way. In the case of polythene the molecule M, which in this context is called a *monomer*, is ethylene, $H_2C=CH_2$. The outcome of the polymerization is a long chain, a *polymer*, of hundreds of $-CH_2CH_2-$ units. Chemists have found that by starting with different versions of ethylene, such as $H_2C=CHX$, where X can be a group of atoms, they can form polymers with a wide range of properties. Thus, when X is a benzene ring, the polymer is polystyrene and when X is a chlorine atom the polymer is PVC. To obtain the non-stick Teflon® all the hydrogen atoms are replaced by fluorine atoms; that is why its more formal name is polytetrafluoroethylene (PTFE).

Another variety of acids: Lewis acids

The fourth and final type of fundamental reaction might seem arcane at first sight, but it is a seriously important process. We have just seen that two radicals can form a bond with each other if each brings along one electron, which then pair. In this final type of reaction, one molecule provides *both* electrons of the bond that forms between them, the other partner in the reaction accommodating both electrons. We could represent this sort of reaction by A + :B → A–B, where the double dot on B represents the electron pair that is en route to be shared with A. Reactions of this kind are called *Lewis acid–base reactions* after the American chemist G. N. Lewis, who first identified them and was later killed by them. (He died after ingesting cyanide ions, CN^-, a poison that acts by this kind of reaction). They are called 'acid–base' reactions because they show marked similarities to the acid–base reactions I discussed earlier, in which a proton migrates from an acid to a base. Indeed, they can be regarded as yet another generalization of the concepts of acid and base, but I will not take you down the fascinating scenery of that route.

One role of Lewis acid–base reactions is to bring colour to the world. This remark gives me the opportunity to introduce you to *transition metal complexes*, which are often brightly coloured and which are formed in what I shall call a Lewis way. The haemoglobin of your blood is an example.

A transition metal is one of the elements in the skinny central part of the Periodic Table, and includes iron (Fe), chromium (Cr, this name anticipates the colour to come as *chroma* is the Greek word for colour), cobalt (Co), and nickel (Ni). The ions these elements form, such as Fe^{2+} and Co^{3+}, are commonly found surrounded by and bonded to six small molecules and ions that have an independent existence, such as H_2O, NH_3, and CN^-. These species are called *ligands* and the complete clusters are called *complexes*.

A complex is held together by bonds formed by the sharing of a pair of electrons provided by each ligand, so the metal ion is acting as the Lewis acid (the A) and each ligand is acting as the Lewis base (the :B).

In water, transition metal ions are typically surrounded by six water molecules acting as Lewis bases. When another Lewis base is added to the solution, it might drive out one or more water molecules and take their place. The electronic structure of the resulting complex might be quite different from that of the original complex, and as a result be brightly coloured. Many pigments and dyes are complexes formed in this way.

Breathing is a Lewis acid–base reaction. The oxygen carrier in our blood is haemoglobin, a huge protein molecule that has embedded in it four iron ions. Each one is gripped in place by four nitrogen atoms belonging to the protein framework and lying round it at the corners of a square. The bonds between the iron and the nitrogen atoms are the result of Lewis acid–base interactions, with Fe^{2+} the acid and each :N a base. When you breathe in, this already Lewis-constructed entity takes part in another Lewis acid–base reaction when an oxygen molecule acting as a Lewis base uses a pair of electrons to form a bond to an Fe^{2+} ion in the haemoglobin molecule. Once captured, the precious oxygen is transported in the blood stream to take part in other reactions deep inside our body.

Suffocation by carbon monoxide poisoning is another Lewis acid–base reaction. Now the carbon monoxide molecule, CO, can usurp the place of oxygen and attach Lewis-like to the Fe^{2+} ions in haemoglobin. This attachment is stronger than oxygen is able to achieve, so the usurper blocks the attachment of oxygen, and there is none transported to where it is needed and the victim suffocates. This is suffocation at a molecular level, not just the blocking of an airway. The poisoning effect of the cyanide ion, CN^-, I mentioned earlier is similar, but it blocks a cascade of electron transfer reactions later in the respiration process.

The subtlety of organic chemistry

Organic chemists are magicians, or general officers commanding, when it comes to deploying these fundamental types of reactions. They need to be, because the molecules they aspire to build are often delicate traceries of atoms, and one atom out of place could render a pharmaceutical inactive or set back research for months. Over the decades of development of organic chemistry, chemists have accumulated a fund of experience in knowing how to coax atoms into the appropriate arrangement to suit their need, sometimes in sequences of reactions with dozens of steps, any one of which might reduce a hard-won compound to the chemist's equivalent of rubble, a black, useless tar. The procedures often go by the names of the chemists who developed them. Computer software is also a help in devising strategies, just as it is used for establishing the workflow of a construction project.

The metaphor of a construction project can be taken further. Just as a partially completed component of the project might need to be protected while building goes on around it, so a partially constructed molecule might have tender regions that, without protection, would be centres of reaction and result in unwanted products. Thus, chemists sometimes might attach a small group of atoms to a region of the molecule either to shield a neighbouring region from attack or to conceal the atom to which it is attached. That protective group can later be stripped off, just like the protective shroud of a building.

I shall give just two examples of how organic chemists go about the business of building a molecule, perhaps one destined to be tested as a pharmaceutical, a dye, or an artificial flavour. Both are examples of a *substitution reaction*, in which an atom or group of atoms is substituted for one already present in a molecule. In each case the target atom is detected by the incoming reactant as a region of the molecule that has either a relatively thin or dense electron cloud. If the cloud is thin, the positively charged nuclei shine through, and a negatively charged reactant molecule will home in on it like a guided missile. Reactions of this nuclear-charge-detecting kind are called *nucleophilic substitutions*. If instead the cloud is dense, then the negative charges of the electrons will outweigh the positive charge of the nuclei, and an incoming missile that is positively charged will home in on the region. Such electron-rich-seeking reactions are called *electrophilic substitutions*.

When planning the construction of a molecule, a chemist needs to think about the way that the electron clouds are distributed in a molecule, and then choose the reactant molecule accordingly. They can be very subtle about the procedure, because it is possible to attach groups of atoms that suck electrons away from a region or alternatively push electrons on to it. By modifying the electron cloud in this way, a chemist can be reasonably confident that a reactant molecule will home in on the right atom and form a bond there.

I hope that, at this point, you are beginning to be able to sense the subtlety with which chemists go about the task of creating forms of matter that might not exist anywhere else in the universe. It will not be possible from this necessarily brief account to comprehend the details of how chemists contrive their reactions, but I hope that you will perceive the thoughtfulness behind their activities.

Chapter 5
Its techniques

Go into any modern chemical laboratory and you will find it a hybrid. An alchemist would recognize some of the apparatus; the rest would be wholly alien. There are only so many shapes for vessels to contain fluids and most of them have a clear ancestry in the past. But modern *analysis*, literally the breaking down of substances and in modern practice the identification of substances and the determination of their amounts and concentrations, makes use of sophisticated electronic and often automated equipment. Analysis is not the only pursuit in a laboratory, for its opposite, *synthesis*, literally the putting together but in practice the creation of desired forms of matter from simpler components, is a major component of a chemist's endeavour.

Classical laboratory equipment

I shall not dwell on beakers, flasks, and test-tubes, for their use for containing and mixing fluids is obvious. Some containers, though, are designed to deliver known amounts of liquids either to accord, as in a kitchen, with a precise recipe or as part of a method of quantitative measurement. An example of the latter is the use of a pipette (US: pipet; a little pipe) and a burette (US: buret; a word derived from the French word for a small vase or jug, although except to the imaginative it is nothing like one) in one of the classic procedures of chemistry, the titration of an acid with a base to determine the concentration of one or the other of them. (Why 'titration'? *Titre* is the French word for assay or test). The pipette is used to deliver a fixed amount of the basic solution to a conical flask; the burette is used to dribble in the acid until a colour change or an electronic signal from a detector signals that the base has been exactly neutralized. By noting the volume of acid added from the graduated burette and knowing its concentration, the concentration of the base can be determined.

One other class of apparatus is concerned with the separation of substances, perhaps to purify or isolate a product. One straightforward technique, when the product is a solid that has been precipitated when two solutions are mixed, is 'filtration': passing the resulting solution through a fine mesh. Another, often used when liquids need to be separated, is 'distillation': boiling the liquid mixture and condensing the vapour; the more volatile component of the mixture boils off first, and may be collected or discarded.

One highly sophisticated separation technique is 'chromatography'. This technique was born and named when it consisted of little more than noting that a drop of solution, perhaps taken from a flowering plant, would spread across absorbent paper and form bands of different colours that could then either be identified or collected. The name survives, but the technique has been immeasurably elaborated. Now, in a typical procedure, the sample to be analysed is passed through many metres of narrow tubing, the interior of which is coated with an absorbent solid. The components of the mixture stick to (the technical term is 'adsorb on') the surface to different degrees, and although they all make it to the end of the tube, they emerge at different times and can thus be collected separately and identified by other procedures. This technique is used to separate the myriad compounds that contribute to the flavour of a fruit and, in a more specialized form, to sniff out explosives, such as at security installations.

Spectroscopy

Much more interesting, and bewildering to the alchemist, is the electronic equipment in the room, showing only its screens and dials and not advertising immediately its purpose. Many of these procedures are forms of *spectroscopy*. The term is derived from the Latin word *spectrum*, or appearance, and the process of looking at the appearance; but 'looking' is now more sophisticated than visual inspection and 'appearance' far removed from its everyday meaning.

I shall begin with atomic spectroscopy. When an element is vaporized and heated, one or more of the electrons of an atom may be ejected from its normal distribution and briefly hang above the atom before collapsing back into its normal cloud again. That collapse gives an impulse to what we think of as the vacuum that surrounds the atom, and the impulse generates a pulse of light, a *photon*. The colour of the photon depends on the energy released in the collapse, with high-energy collapse giving a pulse of ultraviolet radiation and lower energy pulses giving visible light. The electrons of atoms can exist in a variety of energy states that are characteristic of the element, and as the electrons collapse from the state that they happen to have been promoted into they generate photons of the corresponding colours. We are all familiar with the yellow of street lighting, which is due to sodium atoms generating photons as they collapse into their normal state, and of red neon signs, which is due to the electron of a neon atom collapsing into its normal state. By noting the pattern of colours, by 'recording the spectrum', the element present can be identified.

The electrons of molecules behave in much the same way, but monitoring their possible energies is carried out in a somewhat different manner. Whereas the atomic spectroscopy that I have described makes use of the emission of light, molecular spectroscopy does the opposite: it makes use of the absorption of light.

Light that is passed through a sample can be thought of as a stream of photons. One of those photons will be absorbed if it collides with a molecule that can be excited into a higher energy state with a matching energy. The removal of such photons, their 'absorption', from the incoming stream will reduce the intensity of the beam, which will be recorded by a detector of some kind. To record the full absorption spectrum, the colour of the incident light is changed systematically, and the intensity that manages to survive passage through the sample is monitored. Because molecules have characteristic energy levels, their absorption spectra are unique, and can give a good indication of their identity.

I have focused on the absorption of photons by the excitation of electrons from their normal distribution in a molecule. That takes a lot of energy, and although many molecules absorb visible light (which is why the world is so colourful) the spectra I have described are commonly observed by using ultraviolet radiation. Thus, the technique is known as 'UV-vis spectroscopy'. A closely related technique uses photons of infrared radiation, which have much lower energy than visible and ultraviolet photons. These photons can stimulate the vibrations of molecules, not their electron distributions. 'Infrared spectra' therefore show that vibrations can be stimulated. That is very helpful for analysing the groups of atoms present in a complex molecule because a CH_3 group, for instance, can waggle around with one energy and a CO group can waggle around with a different energy.

Nuclear magnetic resonance

Perhaps the single most important analytical spectroscopic technique is 'nuclear magnetic resonance' (NMR). The word 'nuclear' raises a red flag wherever it occurs, which is why it has been dropped from the medical investigatory technique of 'magnetic resonance imaging' (MRI), a technique that is derived from NMR itself. Chemists on the whole are less squeamish than the public at large and retain the word 'nuclear', knowing that in this usage it has nothing at all to do with the perils of radioactivity.

The 'nuclear' of nuclear magnetic resonance refers to any nucleus of any atom, but I shall focus on its most common target, the proton, the nucleus of a hydrogen atom. A proton spins on its axis, like the Earth, and that spinning charge behaves like a tiny bar magnet. It can spin either clockwise or anticlockwise, and the corresponding little bar magnet has its North pole either up or down according to the direction of spin. When the spinning proton is in a magnetic field (in practice, an intense one generated by passing a current through a superconducting coil) the two orientations have different energies, and an incoming photon of the appropriate frequency can flip an upward pointing (low energy) proton into a downward pointing (high energy) proton. The matching of the photon frequency to the energy separation is the 'resonance' of the name; we do the same thing when we tune a radio to the frequency of a distant transmitter. When that takes place, the incoming stream of photons is attenuated, and the decrease in intensity is detected. The energy separation is not very great, and photons of radiofrequency radiation, just off the high-frequency end of an FM radio signal (100 MHz or so), are used.

It might seem a rather pointless activity to flip a proton from one orientation to another. The power of the technique—and that power cannot be underestimated—is that the precise frequency at which resonance occurs depends on where the proton, specifically the hydrogen atom of which it is the nucleus, lies in the molecule. Hydrogen atom nuclei with carbon atoms as neighbours resonate at different frequencies from those with oxygen or nitrogen atoms as neighbours, and so the spectrum of resonant absorptions, the 'NMR spectrum', portrays the neighbourhoods of all the hydrogen atoms in the molecule.

That's not all. The little bar magnets at the heart of hydrogen atoms in the same molecule interact with one another and modify each other's energies. That modification affects the resonant frequencies and gives rise to characteristic patterns of absorption, which is a huge help when trying to identify a molecule.

A carbon nucleus does not spin and so does not behave like a bar magnet and is invisible in NMR. That is a blessing, because otherwise even a quite simple organic molecule would give an impossibly complex NMR spectrum. However, carbon atoms can be revealed cautiously by replacing ordinary carbon atoms, carbon-12, with an isotope, carbon-13, which has an extra neutron in its nucleus and is magnetic. Judicious replacement of carbon-12 by carbon-13 can therefore be used to map the locations of carbon atoms too, and the identity and structure of the molecule can then be pinned down unambiguously.

Mass spectrometry

There is another totally different kind of spectrometer that does not use absorption or emission and gives an entirely different insight into the identity of a molecule. In a 'mass spectrometer' a molecule is blasted apart and its fragments weighed, the composition of the molecule then being inferred from the masses of the fragments.

The molecular shattering is carried out with a blast of electrons, which strike the molecule, distort the electron clouds holding it together, and give rise to a number of electrically charged fragments. These charged fragments are accelerated by an electric field and pass between the poles of a powerful magnet, which bends their paths to an extent that depends on their mass and the strength of the field. Fragments of a particular mass will fall on a detector and give a signal. As the magnetic field is changed, fragments of different masses will come into view by the detector, and the spectrum, now a 'mass spectrum' of masses of fragments, is interpreted in terms of the structure of the parent molecule rather as a smashed vase can be rebuilt from its fragments.

X-ray diffraction

In biology, structure is crucial to function. Structure is almost everything in chemistry, and especially so where chemistry merges with biology and chemists contribute to the study and elucidation of the action of the big protein molecules we know as enzymes. Although enzymes are hugely important for regulating the chemical reactions that constitute all aspects of life, they are not the only crucial components of organisms (which include humans). Inheritance is enabled by DNA, scaffolding by rigid proteins and bone, and perception and thought by molecules that detect and convey messages. The mechanism of the whole body of an organism is modulated by its marinating molecules.

One of the most powerful tools for discovering structure is 'X-ray diffraction' or, because it is always applied to crystals of the substance of interest, 'X-ray crystallography'. The technique has been a gushing fountain of Nobel prizes, starting with Wilhelm Röntgen's discovery of X-rays (awarded in 1901, the first physics prize), then William and his son Laurence Bragg in 1915, Peter Debye in 1936, and continuing with Dorothy Hodgkin (1964), and culminating with Maurice Wilkins (but not Rosalind Franklin) in 1962, which provided the foundation of James Watson's and Francis Crick's formulation of the double-helix structure of DNA, with all its huge implications for understanding inheritance, tackling disease, and capturing criminals (a prize shared with Wilkins in 1962). If there is one technique that is responsible for blending biology into chemistry, then this is it. Another striking feature of this list is that the prize has been awarded in all three scientific categories: chemistry, physics, and physiology and medicine, such is the range of the technique and the illumination it has brought.

To understand the basis of the technique it is essential to know that X-rays are beams of very short wavelength electromagnetic radiation, like light, but with wavelengths a thousand times shorter (around 100 pm (picometres), about the diameter of an atom). The second essential piece of information is that X-rays, like all waves, interfere with each other: where peaks coincide they are brighter; where peaks coincide with troughs, they are dimmer. When an object is put in the path of a beam of X-rays it scatters them, and scattering by different parts of a molecule results in beams that travel to a detector by different paths and so can interfere with each other in a variety of ways. This interference caused by an object in its path is the 'diffraction' part of the name.

In an X-ray diffraction determination, a tiny crystalline sample is rotated in the path of a beam of X-rays, and a detector is made to travel all over a surrounding sphere of locations, detecting the glints of constructive interference as it goes. From that huge number of observations a mathematical trick can be used to establish the arrangement of atoms in the sample. The technique is now extensively automated, with an integrated computer controlling the collection and interpretation of data.

The most challenging part of the determination is making the crystal that is essential to the technique, especially for the big molecules that are one of its principal targets of study. If, however, it is only identification of a substance, such as a mineral, that is required, then it is possible to use a simpler technique in which the sample is a powder spread on a plate. When a beam of X-rays is directed on to the powder, the resulting 'powder diffraction pattern' is characteristic of the substance, and it can be identified by referring to a library of patterns.

Why X-rays? Diffraction patterns are obtained when the wavelength of the radiation is comparable to the scale of the structures that cause it, in our case the atoms, causing the diffraction. It happens that X-ray wavelengths are comparable to atom–atom separations in molecules, so are ideal for the purpose.

Picturing surfaces

The interior of solids is a fascinating place, but the action often takes place on the surface. For instance, in catalysis, the acceleration of reactions by the presence of an otherwise non-participating substance, often occurs through a mechanism involving the attachment of reactants to a solid surface, sometimes ripped apart, where they are rendered ripe for reaction with other reactants. The chemical industry owes its existence to catalysts, so the study of events at surfaces is of great importance.

Surfaces, despite being the outward show of a solid, were hard to study until a few years ago when a dramatic new technique burst on to the scene. This technique is so sensitive that it can portray the individual atoms on the surface and also molecules stuck to it. It comes in two forms: 'scanning tunnelling microscopy' (STM) and 'atomic force microscopy' (AFM).

At first sight, STM might seem rather unlikely. A needle is pulled out to form a very fine point, and then swept in successive rows across the surface being studied. The flow of electric current between the needle point and the surface is monitored and mapped on a screen: atoms that protrude from the surface lie close to the passing needle point and give rise to a surge in current, which is portrayed as a peak on the screen. The success of the method depends on a quantum mechanical effect called 'tunnelling' (hence the tunnelling in the name of the technique), in which electrons are able to cross forbidden regions, in this case the gap between the surface and the needle. Tunnelling is very sensitive to the width of the gap, so the scan across the surface can pick up atom-size variations in the surface itself, and also show up, and depict in compelling detail, the shapes of molecules stuck to the surface.

It is widely claimed that atoms are too small to be seen, however, provided we enlarge our vision of 'seeing' to the graphical portrayal of the variation in tunnelling current, then STM denies that claim and provides us with the most extraordinarily compelling images of individual atoms and molecules. Even the clean surfaces themselves are compelling, with Mars-like mountains and cliffs where atoms pile high and chasms where they have been lost. Surfaces now are open to detailed, direct inspection.

Atomic force microscopy brings direct action to surfaces. Instead of passively observing, the tip is used to move atoms around on the surface by nudging them from place to place. Apart from allowing such entertainments as 'nanosoccer' by moving a C_{60} 'buckyball' over the surface, exquisite control can be exercised on the arrangement of individual atoms. If the tip is coated with molecules of a certain kind, then by moving the tip patterns can be written on to the surface and structures built up on a nanoscale (see Chapter 7).

Computational chemistry

An instrument that has transformed chemistry, just as it has transformed life in general, in recent decades is the computer. Just about all laboratory procedures, except the most primitive, are controlled by computers. As we have just seen, computers are intrinsic to X-ray crystallography, and are essential to the interpretation of diffraction patterns. They are also essential to modern NMR, where special techniques are used to observe the spectrum and need extensive mathematical manipulation to mine for the actual spectrum. There is, however, an application of computers in their own right, that of the computation and graphical portrayal of molecular structures. This is the field of 'computational chemistry'.

Together with weather forecasters and code breakers, chemists are among the most demanding users of powerful computers, although such is progress in computational hardware that much analysis can now also be carried out on a tablet or even a smartphone.

One wing of computational chemistry takes the subject all the way back to the quantum mechanical description of the distribution of electron clouds in molecules, and sets out to calculate those distributions. Such calculations involve a great deal of numerical manipulation and a variety of approximations. Although the output is essentially just a list of numbers relating to the density of the electron cloud throughout the molecule, those numbers are brought to life and rendered digestible by graphical displays of the cloud that enable chemists to assess the likely behaviour of the molecule. One very important application, which is greatly helped by the ability to picture the regions of dense and sparse electron clouds, is in the assessment of the pharmaceutical activity of a molecule and the initial screening of likely pharmacologically active compounds before they are tested *in vivo* on animals.

The second, closely related wing of computational chemistry has to do with how a protein folds into its active shape. A protein molecule is just a long chain of chemically linked molecules (amino acids), yet it folds into helices and sheets and they in turn fold into a reasonably rigid structure that is essential to its function. Although the forces that act between different parts of the same molecule are well understood, it is still an elusive problem to see how all those varied forces conspire to twist the chain into its final shape. Nature does it: we do not yet understand how. As a part of the attack on the problem, computers are used to track the links in the molecular chain as they wriggle and writhe into their final shape in an attempt to understand how Nature does it without so much as a thought.

Computers are used to study the behaviour of little molecules too. One molecule is sent flying towards another (in the imagination of the software), and calculations are used to watch what happens in the most intimate moments of a chemical reaction, when molecules collide, old bonds weaken, and new bonds form.

Modern approaches to synthesis

Now I would like to switch attention away from investigation towards synthesis and to a particular version of synthesis that is currently in vogue, where chemists do not always have a clue about what it is that they have made. I speak of 'combinatorial chemistry'.

The traditional procedure for making compounds is to work on one at a time, with a target clearly in mind. In combinatorial chemistry hundreds, even thousands, are made simultaneously and then examined for the appropriate behaviour, sometimes in similar herds, and when promising candidates have become apparent only they are analysed, their identities determined, and then used as the basis for further work.

The procedure originated with the synthesis of short strips of protein-like molecules known as peptides, and I shall use them to illustrate how it is done. The same strategy has been developed for a range of other types of compound, and has provided an important boost to the throughput of 'drug discovery', the process of formulating medicinally active compounds.

Suppose we have three amino acids, A, B, and C. In Round 1 we prepare a container with A in it, and proceed to carry out a reaction with all three acids, which will result in the compounds AA, AB, and AC. We then bring these three compounds together, mix them up, and divide them into three equal portions, each one containing all three compounds. That is the end of Round 1. Round 2 repeats the process, and in one container, the one that reacts with A, we shall get the three compounds AAA, ABA, and ACA, in the second container, which reacts with B, we get AAB, ABB, and ACB, and similarly for the third container, where we make ACA, ACB, and ACC. These nine compounds are used as the starting point for the next round. In practice, all 20 naturally occurring amino acids might be used rather than the miserly three of this illustration, and four successive rounds give 400, 8,000, 160,000, and 5,200,000 compounds. Thus, millions of compounds can be made (by robots) in almost the twinkling of an eye.

Thus it is that chemists make substances, not always knowing or caring what they have made, and hoping that amid the plethora of products there lie pearls. Clearly, there is a considerable bookkeeping task needed to keep track of the possible substances in each mixture (for instance, in our three-amino-acid example, after the second round the first container has only three candidates present), and a computer keeps track of what the robots are up to. If in some subsequent test the contents of a mixing vessel show a certain biological activity, such as inhibiting a malfunctioning enzyme that is responsible for a disease, then those contents can be candidates for separation and identification, the rest being washed away as of no interest.

There was a stage, only a few years ago, when chemists were proud to have made and identified about 10,000,000 compounds. Now they might make several times that number in a month, and only occasionally bother to determine what they have made. Such is progress.

Chapter 6
Its achievements

I have already remarked that life without chemistry would see us back in the Stone Age. Almost all the infrastructure and comforts of the modern world have emerged from chemical research. In the primitive days of the subject, when curiosity, tradition, and alchemy were collaborators and science was still the merest green shoot, that research was unaided by correct theory and progress was haltingly slow. Now that the subject is mature, with curiosity in fruitful alliance with understanding and exploitation, the research is largely rational and its achievements substantial.

In the broadest possible terms, chemists have discovered how to take one form of matter and conjure from it a different form. In some cases they have discovered how to take raw material from the Earth, such as oil or ore, and to produce materials directly from them, such as petroleum fuels and iron for steel. They have also discovered how to harvest the skies, how to take the nitrogen of the atmosphere and convert it into fertilizer. They have also discovered how to make highly sophisticated forms of matter suitable for use as fabrics or as the substances needed for what currently we regard as high technology, knowing that there is still higher technology to come and which will be enabled, we can be confident, by chemistry.

Earth, air, fire, and water

I shall begin this account of chemistry's numerous achievements by considering the famous four so-called elements of antiquity: earth, air, fire, and water.

First water, the absolutely essential enabler of life both at the level of individual organisms and at the level of global societies. Chemistry has made communal living possible through its use to purify water and rid it of pathogens. Chlorine is the principal agent for enabling cities to exist: without it, disease would be rampant and urban living a gamble and more akin to urban dying, just as it used to be. Chemists have found ways of extracting this element from an abundant source—sodium chloride, common salt—by using electrolysis to oxidize the chloride ions in the molten salt and by stripping off an electron from each one, so converting them to the element. The virile gas chlorine then goes on to attack the pathogens, rendering them harmless.

Chemists are at the forefront of the battle to obtain potable water from brackish water, from poisoned water in aquifers, and from that most abundant source of all, the oceans. They have contributed in direct ways to this crucial task by developing 'reverse osmosis', the process in which water is squeezed through membranes that filter out the ions that render it undrinkable. They contribute too in indirect ways by developing the membranes that can withstand the high pressures involved and contribute to the efficiency of the process. It goes without saying that chemists' traditional skills of analysis, discovering what is present, what can be tolerated, and what it is essential to remove, are crucial to this endeavour.

Then take earth, food's source. As the global population grows and the productive land area is eroded, so it becomes more and more important to coax crops into greater abundance and fecundity. Genetic engineering (a chemical technique, performed literally in fruitful collaboration with biology) is one way to proceed, but remains controversial for a variety of reasons, some plausible, others not. The traditional way to encourage abundance is to apply fertilizers. Here, chemists have contributed substantially by finding economically viable sources of nitrogen and phosphorus and ensuring that they can be converted to a form that can be assimilated by plants.

Air supplies the earth. Nitrogen (N), one of the elements essential to agriculture, is astonishingly abundant, making up nearly three quarters of the atmosphere; but it is there in a form that cannot be assimilated by most plants. This stubborn inertness is due almost entirely to the fact that the two nitrogen atoms of a nitrogen molecule (N_2) are strapped together by a powerful triple bond, three shared pairs of electrons, and are notoriously difficult to separate. Indeed, that is the principal reason why atmospheric nitrogen is so abundant in the air: it simply remains aloof from most attempts to react with it, requiring lightning bolts or the bacteria associated with certain leguminous plants to trap it chemically.

One of chemistry's greatest achievements, made in the opening years of the 20th century under the impetus not of a humane desire to feed but of an inhumane desire to kill, was to discover how to harvest nitrogen from the air and turn it into a form that could be absorbed by crops (and used to make explosives). This achievement, by Fritz Haber and Carl Bosch, was a landmark in the chemical industry, for as well as depending on the discovery of appropriate catalysts to facilitate the reaction between nitrogen and hydrogen gases to form ammonia (NH_3), it required the development of industrial plant that operated at temperatures and pressures never previously attained. But the process, which is still used globally today, remains energy intensive. It would be wonderful if the processes known to occur in the bacteria that inhabit the nodules of the roots of alfalfa, clover, beans, peas, and other legumes could be emulated on an industrial scale and harnessed to harvest atmospheric nitrogen. Chemists have put decades of research into this possibility, dissecting in detail the enzymes that bacteria use in their quiet and energy-efficient, low pressure, low temperature way. There are glimmerings of success, but no method is yet commercially viable.

Phosphorus (P) is abundant too, being the remains of prehistoric animals. Their bones of calcium phosphate and their special internal power source, the molecules of ATP (adenosine triphosphate) that power every one of our and their cells, lie in great compressed heaps below the oceans of the world as phosphate rock. Here chemists help to mine the dead to feed the living, for they find ways to extract the phosphorus from these buried sources and use it again in the great cycle of sustainability.

After water, air, and the food that springs from the earth, we need energy, the representative of fire in this quartet. Nothing happens in the world without energy, and civilizations would collapse if it ceased to be available. Civilizations advance by deploying energy in ever greater quantities, and chemists contribute at all levels and in all aspects of the development of new sources and more efficient applications of current sources.

Petroleum is, of course, an extraordinarily convenient source of energy, as it can be transported easily, even in weight-sensitive aircraft. Chemists have long contributed to the refinement of the raw material squeezed and pumped from the ground. They have developed processes and catalysts that have taken the molecules provided by Nature and used them to cut the molecules into more volatile fragments and reshape them so that they burn more efficiently. But burning Nature's underground bounty might by future generations be seen as the wanton destruction of an invaluable resource, akin to species extinction. It is also finite, and although economically viable new sources of petroleum are constantly, for the time being at least, being discovered, it is proving hazardous and increasingly expensive to extract it. We have to accept that although an empty Earth is decades off, one day it will arrive and needs to be anticipated.

Where do chemists currently look for new sources? The Sun, that distant, furious, nuclear fusion furnace in the sky, is an obvious source, and the capture of its energy that Nature has adopted, namely photosynthesis, is an obvious model to try to emulate. Chemists have already developed moderately efficient photovoltaic materials, and continue to develop their efficiency. Nature, with her 4 billion year start on laboratory chemists, has already developed a highly efficient system based on chlorophyll, and although the principal features of the process are understood, a challenge for chemists is to take Nature's model and adapt it to an industrial scale. One route is to use sunlight to split water into its component elements, desirable hydrogen and already abundant oxygen, and to pipe or pump the hydrogen to where it can be burned.

I say 'burned'. Chemists know that there are more subtle and efficient ways of using the energy that hydrogen and hydrocarbons represent than igniting them, capturing the energy released as heat, and using that heat in a mechanical, inefficient engine or electrical generator. *Electrochemistry*, the use of chemical reactions to generate electricity and the use of electricity to bring about chemical change, is potentially of huge importance to the world. Chemists have already helped to produce the mobile sources, the batteries, that drive our small portable devices, such as lamps, music players, laptops, telephones, monitoring devices of all kinds, and increasingly our vehicles.

Chemists are deeply involved in collaboration with engineers in the development of 'fuel cells' on all scales, from driving laptops to powering entire homes and conceivably villages. In a fuel cell, electricity is generated by allowing chemical reactions to dump and extract electrons into and from conducting surfaces while fuel, either hydrogen or hydrocarbons, is supplied from outside. The viability of a fuel cell depends crucially on the nature of the surfaces where the reactions take place and the medium in which they are immersed.

Even nuclear power, both fission and one day fusion (the emulation on Earth of the Sun), depend on the skills of chemists. The construction of nuclear reactors depends on the availability of new materials, and the extraction of nuclear fuel in the form of uranium and its oxides from its ores involves chemistry. Everyone knows that one fear that holds back the development and public acceptance of nuclear energy, apart from political and economic problems, is the problem of how to dispose of the highly radioactive spent fuel. Chemists contribute by finding ways to extract useful isotopes from nuclear waste and by finding ways to ensure that it does not enter the environment and become a hazard for centuries.

Artefacts from oil

I have alluded to the seemingly wanton destruction of an invaluable resource when the complex organic mixture we know as oil is sucked from the ground where it has lain for millennia and then casually burned. Of course, not all the oil is used in engines and its combustion products spewed out through the exhausts of our cars, trucks, trains, and aircraft. Much is extracted and used as the head of an awesome chain of reactions that chemists have developed and which constitute the petrochemical industry.

Look around you and identify what chemists have achieved by taking the black, viscous crude oil that emerges from the Earth, subjecting it to the reactions that they have developed, and passing on the products to the manufacturers of the artefacts of the modern world.

Perhaps the greatest impact of these processes has been the development of plastics. A century ago the everyday world was metallic, ceramic, or natural, with objects built from wood, wool, cotton, and silk. Today, an abundance of objects are built from synthetics derived from oil. Our fabrics have been spun from materials developed by chemists, we travel carting bags and cases formed from synthetics; our electronic equipment, our televisions, telephones, and laptops are all moulded from synthetics. Our vehicles are increasingly fabricated from synthetics. Even the look and feel of the world is now different from what it was a hundred years ago: touch an object today, and its texture is typically that of a synthetic material. For this transformation, we are in debt to chemists who have discovered how to chop up the long molecules exuded by the Earth and then reassemble them into very long chains in the process of polymerization. Thus, ethylene ($CH_2 = CHX$, with $X = H$) is spun into polyethylene and used for everything from shopping (as plastic bags, a mixed blessing) to helping win World War II (as cladding for radar cables). As I mentioned in Chapter 4, when X is chlorine, Cl, the monomers are spun into PVC, which has taken over from wood and metal in much construction work.

Although the convenience of plastic bags is perhaps outweighed by their blight on the environment, think of what we would not have if we had none of the polymeric materials invented by chemists and then fabricated in bulk. Think of a world without nylon and the polyesters of fabrics for clothing, upholstery, and decoration. Think of a world with only heavy metal containers for drinks, food, and household fluids. Think of a world without all the little plastic artefacts of everyday life, switches, plugs, sockets, toys, knife handles, keyboards, buttons... the list is almost literally endless, so ubiquitous is the presence of chemistry-generated polymeric materials.

Even if you mourn the passing of many natural materials, you can still thank chemists for their preservation where they are still employed. Natural matter rots, but chemists have developed materials that ensure that that decay is postponed. In short, chemists both provide new materials when those are judged appropriate or desirable, and provide means of prolonging the lives of natural materials when judgement and choice leads to their adoption.

Plastics are but one face of the revolution in materials that has characterized the last one hundred years and is continuing vigorously today. Chemists develop the ceramics that are beginning to replace the metals that we use in vehicles, so lightening them and increasing the efficiency of our transport systems with the consequent lessening of its impact on the environment. Ceramics, of course, are materials of great antiquity, for they are the stuff of pots (another largely unacknowledged contribution to the viability of social life). Modern ceramics are tailored more systematically from purified clay and other materials, and sometimes exhibit surprising properties. Who, for instance, would have suspected that one class of ceramics baked from an almost witch's brew of elements would have possessed the remarkable property of *superconductivity*, the ability to conduct electricity without resistance? This material, which operates at very low temperatures, but at much higher temperatures than the previously known superconducting materials and therefore more economically encouraging and acceptable, is still groping for applications, for fabricating wires and films from ceramics remains a challenging problem.

Ceramics include glass. Modern glass includes the optical fibre that constitutes the spine of our global communication system. Glass is fundamentally silica (silicon dioxide, SiO_2) from sand that has been purified, rendered molten, and then allowed to cool. Over the centuries chemists have fiddled with this fundamental composition and have given us the richly coloured 'stained' glass, where enthralling hues are caused by impurities added cautiously and selectively. Certainly in the early days the colours were developed by the skill and wisdom of glassmakers, then not specifically chemists. But it is now chemists who formulate the composition of glasses that in some cases are richly coloured but in others, for fibres in particular, are strikingly transparent and able to convey pulses of light over great distances with minimal attenuation.

The creation of colour

The world of human fabrication would be drab without the contributions of chemists. Vibrant colours were once the domain of the wealthy who could afford the expense of purchasing natural colours, such as Tyrian purple, extracted from the glandular mucus of certain sea-snails (*Bolinus brandaris*) where 12,000 snails are milked or wantonly squashed to derive little more than a gram of dye, barely enough to dye the hem of a cloak, or of lapis lazuli (the 'stone of heaven') from distant Afghanistan for deep appealing ultramarine. Then along came William Perkin (1838–1907) who, when attempting unsuccessfully to synthesize quinine, without the advantage of knowing its structure, in an aim to save the empire's armies and bureaucrats from malaria, stumbled instead on the dye he called mauveine, thereby saving sea-snails instead of soldiers from slaughter and incidentally founding the British chemical industry. Thus he laid the foundation for the generation of all his personal and much of Britain's national wealth.

Chemists have added a whole spectrum of colours to the material world, which is no longer drab, except when needed (as in camouflage), but instead can be anything from vibrantly assertive or demurely subtle. Not only is the range of colours now enormous, with fluorescence and reflective sparkle added to the range, but the colours are lightfast and can withstand the rigours of the laundry.

Chemically created colours are not confined to cloth. Pigments in general have been developed; not only the colouring materials themselves but also the support medium, as in the paints used in buildings and the acrylics used by artists. Think of the advances made in household paints, with improvements to their flow properties, their stability in aggressive atmospheres, and their range of colours, including colours that intentionally fade to show where paint is being applied.

Even the colours of television screens and computer monitors make use of solids that have been developed by chemists. Gone are the days of power-hungry, bulky cathode-ray tubes. Now we are in the world of liquid crystals, plasma displays, and OLEDs (organic light-emitting diodes). The liquid crystals and OLEDs are formed of molecules built by chemists that respond in special ways to electric fields and have made possible portable devices with visual displays.

The infrastructure of the everyday

Chemists are also responsible for developing the semiconductors that underlie the modern world of communication and computation. Indeed, one of the principal contributions of chemistry is currently the development of what could be regarded as the material infrastructure of the digital world. Chemists develop the semiconductors that lie at the heart of computation and the optical fibres that are increasingly replacing copper for the transmission of signals. The displays that act as interfaces with the human visual system are a result of the development of materials by chemists.

Currently chemists are developing molecular computers, in which switches and memories are based on changes in the shape of molecules. The successful development of such materials—and with the optimism so typical of science we can be confident that the endeavour will be successful—will result in an unprecedented increase in computational power and an astonishing compactness. If you are interested in the development of such smart materials, then you can expect to contribute to a revolution in computation. There is also the prospect of the development of quantum computing, which will depend on chemists being able to develop appropriate new materials and will result in an almost unforeseeable revolution in communication and computation.

Medicinal chemistry

I have barely mentioned health. One of the great contributions of chemistry to human civilization (and, it must be added, to the welfare of herds) has been the development of pharmaceuticals. Chemists can be justly proud of their contribution to the development of agents that fight disease. Perhaps their most welcome contribution has been the development of anaesthetics and the consequent amelioration of the prospect of pain. Think of undergoing an amputation 200 years ago, with only brandy and gritted teeth to sustain you! Next in importance has been the development, by chemists, often by observing Nature closely, of antibiotics. A century ago, bacterial infection was a deadly prospect, but now, through the availability of penicillin and its chemically modified descendants, it is curable. We have to hope that it remains that way, but we need to prepare for the opposite as bacteria evolve to evade their nemesis.

The pharmaceutical companies often come under attack for what many regard as their profligate profits and exploitation. But they deserve cautious sympathy. Their underlying motive is the admirable aim (albeit with an eye on profit) of reducing human suffering by developing drugs that combat disease. Chemists are at the heart of this endeavour. It is highly regrettable that the development is so expensive. Modern computational techniques are helping in the search for new lines of approach and helping to reduce reliance on *in vivo* animal testing, but extraordinary care needs to be exercised when introducing foreign materials into living human bodies, and years of costly research can suddenly be ruined if at the last stage of testing unacceptable consequences are discovered.

Closely allied with the contribution of chemists to the alleviation of disease is their involvement at a molecular level. Biology became chemistry half a century ago when the structure of DNA was discovered (in 1953). Molecular biology, which in large measure has sprung from that discovery, is chemistry applied to the functioning of organisms. Chemists, often disguised as molecular biologists, have opened the door to understanding life and its principal characteristic, inheritance, at a most fundamental level, and have thereby opened up great regions of the molecular world to rational investigation. They have also transformed forensic medicine, brought criminals to justice, and transformed anthropology.

The shift of chemistry's attention to the processes of life has come at a time when the traditional branches of chemistry—organic, inorganic, and physical—have reached a stage of considerable maturity and are ready to tackle the awesomely complex network of processes going on inside organisms: human bodies in particular. The approach to the treatment, more importantly the prevention, of disease has been put on a rational basis by the discoveries that chemists continue to make. If you plan to enter this field, then genomics and proteomics will turn out to be of crucial importance to your work. This is truly a region of chemistry where you can feel confident about standing on the shoulders of the giants who have preceded you and know that you are attacking disease at its roots.

Warfare, and other evils

Then there is the dark side of chemistry. It would be inappropriate in this account of chemistry's great achievements for no mention to be made of its ability to enhance humanity's ability to damage and kill, for those achievements have come at a cost, in some cases to human life, in others to the environment.

First, the advances made in killing and maiming. Chemists have been responsible for the development of gases for warfare and the optimization of explosives. Indeed, Fritz Haber, whom I have mentioned in connection with his invention of the process of the synthesis of ammonia that has led to the widespread availability of potent fertilizers, was also a leader in the development of poison gas. There is the hope that the elimination of such weapons will enable us to judge his net contribution to human life more kindly, despite how we judge his personality. Although governments have the responsibility for using such terrible weapons, the chemists who contributed to their development cannot, in my view, avoid our condemnation. No good has come from the development of chemical weapons that might be put in the opposite scale to mitigate our condemnation of them: they are pure evil. Numerous states, not all the most powerful but covering about 98 per cent of the world's population, have rejected them as illegal weapons of warfare, and it is to be hoped that the rest will follow suit and join the treaty banning them.

Chemical warfare can be waged by accident. Such was the case at Bhopal, India, in 1984, when the Union Carbide plant there ran out of control with the result, according to official sources, of nearly 4,000 deaths directly related to the disaster and a further 8,000 within two weeks, and with over 500,000 injured. Intentional chemical warfare has never been so successful. The proximate cause of the disaster was the entry of water into an over-stocked, under-cooled tank of the compound methyl isocyanate (CH_3NCO), an intermediate in the manufacturing process of a pesticide. The receding demand for the pesticide at the time had resulted in the accumulation of more than normal quantities of the intermediate. How the water entered remains disputed: the company maintains it was sabotage by a disgruntled employee; others maintain that it entered accidentally in a plant where the safety controls were disorganized, ineffectual, missing, inadequate, and disregarded. The ensuing reaction released 30 tonnes of toxic gas into the atmosphere, visiting death and incalculable physical and emotional suffering on the inhabitants of the densely populated surrounding shanty town.

Comments on the inherent dangers of chemical plants would be otiose, and suggesting that the risks outweigh the advantages would be banal. Only very rarely, however, do such catastrophes occur, and we have to hope that lessons learned from the awful price paid will instil better practice in design and operation of the plants that, in the main, contribute to our well-being.

The other dark facet of chemistry is its provision, improvement, and manufacture of explosives. Here the facet is not entirely black, for explosives are useful in quarrying and mining. The black facet is their use in bombs and in the provision of the impelling force of projectiles: bullets, mortars, and the like. Explosives are compounds that when detonated undergo a very fast reaction—essentially, the molecules fragment into tiny pieces that form a gas and the very fast generation of gas creates the destructive or impulsive shock of the explosion.

In the early days of explosives, gunpowder was king. Its action depends on the intimate intermingling of oxidizing agents (sulfur, potassium nitrate) and stuff that can be oxidized (charcoal, essentially an impure form of the element carbon). The migration of the electrons to the oxidizing agents from the carbon, dragging across atoms, results in a large number of little molecules, a gas. Since then, substances and mixtures have been developed that react more rapidly and accordingly give a sharper shock. Instead of the mingling of different components, chemists have worked towards the ultimate intimacy: ensuring that the oxidizing and oxidizable components are parts of the same molecule so that electron transfer and the ensuing atom rearrangement and molecular fragmentation are as fast as possible and that large numbers of small fragment molecules are formed to amplify the shock. Famous among such compounds is nitroglycerin. This highly unstable compound was tamed when Alfred Nobel (1833–96) discovered that it could be absorbed into a type of porous clay, so forming dynamite and in due course providing the funds for the establishment of one of the greatest conscience-appeasing foundations, the Nobel Foundation, committed as its prizes are to the enhancement of the human condition and the propagation of peace.

Environmental issues

While we are in this embarrassingly negative corner of chemistry, I cannot avoid that other great pointed finger, the one directed at the environmental damage laid at the subject's door, or at least at its drains. It is impossible to deny that the unwanted effluent of the chemical plant has wrought ecological havoc. Ever since Perkin's factories turned the nearby canals red, green, and yellow according to the manufacturing priorities of the day, mankind's aspiration for its own betterment has been at an environmental cost. In fact, the green shoots of environmental pollution, if that is not too ironical a term, can be traced back to the Greeks and Romans, for analysis of ice cores laid down in those eras show traces of the consequences of metal working.

The way forward is either legal or chemical. The legal constrains by the prospect of punishment; the chemical avoids by elimination at source. The latter, always the better mode of action, depends on developments of chemistry itself and has inspired the politico-environmento-chemical movement of *green chemistry*. In broad terms, green chemistry aims to minimize the impact of chemical manufacturing processes on the environment by strict guidelines about the use of materials and the elimination of waste.

The protagonists of green chemistry begin with the plausible proposition that it is better to prevent waste than to clean it up after it has been generated. The implication of that fundamental principle is that whatever is used as starting materials in a process should appear in (as close as it is possible to) its entirety in the final product: whatever atoms go in should all appear in the molecules of the product, with as few as possible discarded as unwanted. It is in this implication that there are considerable economic and technological impacts, and therefore commercial reluctance, for processes and plants need to be designed accordingly and specific raw materials acquired from inconvenient sources, possibly at great expense.

With the process optimized, and particularly if the optimization is beyond technological and economic grasp, the procedures should be designed to avoid or at least minimize the involvement, not just as waste but also as potentially escapable intermediates, of toxic compounds. That requirement is also required of the final product, which should offer minimum risk of toxicity for human life (as the formalizers of the principles identified, but it seems more than appropriate to add organisms in general) and the environment. The restraint also applies to auxiliary materials employed in the process, particularly the liquids that are used as solvents and might, perhaps 'might' for some current processes becomes 'must', be released into the environment, even in small amounts, as leaks develop in the recycling procedures. Chemists, even for their own miniscule laboratory procedures, are essential sniffers out of benign solvents and the development of reactions that take place in these unfamiliar novel environments.

Another ideal aspiration of the proponents of green chemistry is that the feedstock should be renewable. Renewability can take a variety of forms, but all avoid the gouging out of resources from the Earth. Nature furnishes crops each year, and they count as renewable due to the benevolence of the Sun and its powering of the recycling of carbon dioxide through the medium of photosynthesis. Materials other than carbon dioxide can be recycled and plans have been proposed for the treatment of landfill as mines, but that resource is hazardous and not open to geological judgement.

The proponents of green chemistry recognize another contribution to waste and pollution: the role of energy in a chemical process. All requirements of energy make demands on the environment, either through the requirement of fuel or the impact of the exhaust on the atmosphere. Ideally, all procedures should take place without the need to heat and, even more expensively and destructively, cool.

Then there are a number of more technical requirements for the process to be as green as possible. Many procedures in organic chemistry, as in the fabrication of pharmaceuticals, require intermediate steps in which molecules are modified temporarily on their way to becoming the final product. Each step needs special conditions, its own reagents, and perhaps a variety of noxious solvents. The procedure shifts towards the green end of the production spectrum by minimizing these intermediates and looking for more direct routes from feedstock to product.

Bright green chemists look beyond the process itself to the whole lifetime span of its product and look for ways to ensure that at the end of its functional lifetime the product and anything into which it decays will not be toxic or degrade while in the environment into toxic remnants. The 'whole lifetime' consideration includes the anticipation of disaster during the manufacturing process itself (recalling Bhopal), with the precautionary implication that whatever is produced or stored should, in the event of accident, have a minimal effect on the environment. The mitigation of the possibility of catastrophe entails the ceaseless and reliable analysis of all the components and conditions of reaction and storage vessels and fail-safe monitoring procedures that cannot, as at Bhopal, be ignored or circumvented.

Such are the aspirations of green chemistry. The underlying consideration is that it is essential to appeal to chemistry to solve, and preferably avoid, the problems it might cause. There is always, of course, a tension between commercial profit and social and environmental responsibility, this tension not being helped by low levels of supervision in some environments which allows industry to get away, almost literally, with murder.

Pandora's box has always been thus: meddling with Nature invariably entails risk. Chemists meddle at the very roots of material Nature, taking the atoms she provides and recasting them into compounds that are alien to her and which, intruding into her ecosystem, can upset the fine balances of life. With this Merlin-like ability to conjure with atoms come responsibilities, which have not always been recognized in the past, but under social pressures are now high in the chemical industry's awareness of its responsibilities.

The crucial consideration, however, is where reliable solutions to the world's problems will come from if it is not further development of chemistry. Chemistry holds the key to the enhancement of almost every aspect of our daily lives, from the cradle to the grave and all points in between. It has provided the material foundation of all our comforts, not only in health but in illness too, and there is no reason to suppose that it has reached its zenith. It contributes to our communications, both virtual and physical, for it provides the materials along which our electrons and photons travel in the complex network of patterns and interactions that result in computation. Moreover, it develops our fuels, rendering them more efficiently combustible and through catalysis minimizing their noxious products, and helps in the migration from fossil fuels to renewable sources, such as in the development of photovoltaic substances. Chemistry is the only solution to the problems it causes in the environment, be it in earth, air, or water.

The cultural contributions of chemistry

There is another achievement of chemistry that it would be inappropriate to ignore in this survey: that it gives insight into the workings of the material world, insights that range from rock to organism. Insights are an enhancement of the human condition, for they lend understanding to wonder and thereby add to our delight.

Through chemistry we understand the composition and structure of the minerals that constitute the landscape and can see into the structures of rocks and know why they are rigid, why they might glisten, why they might fracture and erode, and what they contain. We know why metals can be beaten into shape and drawn into wires, and through our knowledge of the arrangement of their atoms why some bend to our will but others snap. We understand the play that may be made with the properties of metals by forming alloys and steels. We understand the colours of gemstones and why we can see through glass but not through wood.

Through chemistry we can unravel and comprehend the once inscrutable mysteries of the natural world. We can understand the green of a leaf and the red of a rose, the fragrance of a herb and new-mown hay. We can understand, in a halting but increasing way, the intricate and complex reticulation of processes in the natural world that constitute the awesome and multifaceted property we know as life. We are beginning, even more haltingly, to understand the chemical processes in our brains that enable us to perceive, wonder, and understand.

Although chemistry does not deal with the ultimate fabric of the material world, the zoo of fundamental particles that lie in the domain of fundamental particle physics, it deals with combinations of them, atoms, that have distinguishing personalities. Through chemistry we have come to understand the personalities of the elements, understanding why they have these personalities through the structures of their atoms and why they enter into certain combinations but not others. Through chemistry, the very stuff of chemistry, we know how to make use of these personalities to build molecules and forms of matter that might not exist anywhere else in our galaxy.

We understand, through chemistry, the flavours of foods, the colours of fabrics, the texture of matter, the wetness of water, the changing colours of foliage in spring, summer, and autumn. Not every moment of our lives do we need to turn on understanding, for lying back in animal delight can be a pleasure of its own, just basking in the pleasure of our surroundings. But chemistry adds a depth to this delight, for when the mood moves us and the inclination impels, we can look beneath the superficial pleasures of the world and enjoy the knowledge that we know how things are.

Chapter 7
Its future

New elements go on being discovered, currently at the rate of one every year or so, meaning that the Periodic Table is getting bigger with more scope, in principle, for chemists to explore. Unfortunately, all these new elements are multiply useless: they are radioactive and so unstable that they vanish within fractions of a second. Moreover, no more than a few atoms of them are ever made, and immediately vanish in a puff of fundamental particles.

The edge of the unknown

There are theoretical reasons for suspecting that just a little further along in the Periodic Table, at the yet-to-be-made elements numbered about 126 (in 2013 we are up to 116, livermorium, with one or two others as yet unnamed and their sighting not yet confirmed but hinted at through the mists) that they will form what is known as an 'island of stability' and survive for significantly longer than those around them. It is unlikely, though, that any of them will have any useful applications except as test-beds for theories of nuclear structure. Chemists have no reason to think that they will provide an impetus to chemistry.

They have plenty of elements to get on with. New techniques are being developed that promise to extend the sensitivity, precision, and scope of observations. The ability to detect extraordinarily small quantities of materials is both a blessing and a curse. To understand the composition of a sample in exquisite detail brings understanding closer, to detect the hint that bombs have been in a terrorist's hands helps us to survive, but to find contaminants everywhere, for in this ever churning world that will ever be so, can just confuse and perhaps unnecessarily alarm.

New worlds

Important techniques that are being developed include those where small numbers of atoms and molecules can be studied rather than having to infer their behaviour from observations on bulk samples. Chemists want to know the intimacies of molecular interaction and transformation, and being able to examine the properties of molecules in isolation or as they come together and react, with bonds loosening, atoms shaking free, and falling into new arrangements, is the holy grail of chemistry (of physical chemistry, at least). For some years now it has been possible to watch molecules evolve on timescales of the order of femtoseconds (10^{-15} s, 1/1,000,000,000,000,000 s) and progress has been made to extend that scale to attoseconds (10^{-18} s, a thousand times shorter), when even electrons are frozen in motion and chemistry has finally become physics.

When we encounter tiny groups of atoms, interesting questions and special rules come into play. Take water, for instance: what is the smallest possible ice cube? It has been discovered that you need at least 275 water molecules in a cluster before it can show ice-like properties, with about 475 molecules before it becomes truly ice. That is a cube with about eight H_2O molecules along each edge. The importance of this kind of knowledge is that it helps us model the process of cloud formation in the atmosphere as well as understand how liquids freeze.

When dealing with tiny collections of atoms at low temperatures we have to accept that their behaviour is governed by quantum mechanics and that we should expect weird properties. All matter, including everyday matter, is also governed by quantum mechanics, but we deal with such vast numbers of atoms even in a pinch of salt that the weirdness is washed away and we perceive only averages, the familiarities of behaviour of common matter. These new states of matter that are starting to be made might have consequences of little importance for chemistry, but perhaps not: they might be perfect for storing data and for the development of quantum computing.

Chemists are contributing hugely to one emerging field where small numbers of molecules are present: the world of nanoscience and nanotechnology. Nanosystems (from the Greek *nanos*, a dwarf) are composed of entities about 100 nm (10^{-7} m, 1/10,000 of a millimetre) in diameter, and lie in the intermediate region between individual molecules (about a thousand times smaller) and bulk matter (about a thousand times larger). Frontiers are always fascinating places, and this notional frontier between the big and the small is no exception. The nanoparticles (notice how the prefix can be affixed to many nouns: there are more to come) are small enough for quantum effects to be relevant and for thermodynamics, once regarded as a finished theory, to be bewildered and in need of reconsideration.

Here is fruitful ground for physical chemists to explore, and to formulate and refine their conventional theories for application to these hitherto unconventional materials. Here too is where both organic and inorganic chemists have much to contribute, particularly in the fabrication of nanomaterials, for both organic and inorganic substances can be formulated to inhabit the nanoworld. Fabrication can be 'top-down', when nanostructures are carved out of macroscopic materials, like a sculptor at work on marble, or it can be 'bottom-up', when the nanostructures are built up brick by brick. The latter is particularly interesting, as the construction typically takes place by 'self-assembly'. In this hands-off procedure, molecules are constructed that, when shaken together, aggregate into the desired nanostructure, rather as we all might once have hoped that shaking a jigsaw puzzle would assemble the picture as the pieces interlocked spontaneously rather than going through the irksome business of linking them piece by piece by hand.

Nanotechnology, the development and application of nanomaterials, and nanoscience, their study in general, is currently all the rage in chemistry, and rightly so, for nanomaterials hold great promise. Whole institutes are being dedicated to their study. The potential applications of nanomaterials range across disciplines and are already central to many practical applications. For instance, they show superior light-harvesting characteristics compared to traditional silicon solar cells and have been incorporated into sensors for glucose in blood. Materials containing cadmium have been investigated extensively in the latter connection, with fears that the toxic element cadmium might be inappropriate to inject into human bodies; but recent results on primates seem to mitigate this fear. Nanorods, nanowires, nanofibres, nanowhiskers, nanobelts, and nanotubes have also been created, with potential applications in nanomachinery and nanocomputers.

Chemistry is preparing itself to play a major role in the miniaturization of computation. We have seen the impact of the reduction in size (and power consumption) between the early room-filling computers of the 1950s to the tiny, ubiquitous, powerful computers of today and their impact on society and daily life. That was a step from the scale of metres to centimetres, a hundredfold decrease in linear dimension and a millionfold decrease in volume and weight, scaling from room-size to pocket-size but accompanied by a huge increase in computational power. That decrease in size allied with an increase in capability and consequential increase in social impact can be repeated if current progress with the development of molecular computation bears fruit.

Computational procedures depend on two features: memory and manipulation. Memory is quite easy to achieve at a molecular level by causing a molecule to undergo a change of shape that is preserved and accessible to some kind of observation. For instance, a molecule might be caused to bend into a certain shape to represent 1 and bend into a different shape to represent 0. A variety of conformational changes are now available, such as a ring-like molecule sliding to either end of a rod-shaped molecule and staying there. Manipulation is more difficult, but comes down to achieving a certain output from a certain input. Chemistry, though, is all about outputs from inputs in the form of chemical reactions, including the output of light when two reagents meet.

Nature has already solved the problem of data storage in her development of DNA, and has evolved methods of extracting that information and turning it into organisms. Our memories are chemically encoded in as yet unknown ways in the brain and provide an immense but fragile and imperfectly stored database. DNA molecules have been used to perform simple arithmetical operations and to 'decide' the treatment necessary if they encounter a damaged protein molecule. Growing computers rather than making them is still science fiction, but there are hints of it on the horizon.

New dimensions

One remarkable recent development has been chemistry's migration from three to two dimensions. The common pencil-filling material graphite is a form of the element carbon in which the carbon atoms form flat sheets like chicken wire that, when impurities are present, slide over each other perhaps to be left as a mark on a page or to act as a lubricant. The individual sheets are called *graphene*, and the fact that they can be plucked off solid graphite by a very simple procedure helped to earn Andre Geim and Konstantin Novoselov the 2010 Nobel Prize (for physics).

Graphene itself is currently viewed as a great prize for physicists and potentially for engineers. It is one of the strongest materials known, with a breaking point 200 times greater than that of steel, yet is very light, weighing less than a gram per square metre. In the Nobel citation it is remarked that 'a 1 square metre hammock would support a 4 kg cat but would weigh only as much as one of the cat's whiskers'. Its extraordinary electronic, thermal, and optical properties are also of great interest, with among other potential applications the creation of loudspeakers with no moving parts and which can be moulded to different surfaces, and achieving in effect the room-temperature distillation of vodka, essentially filtering off the water.

Where do chemists stand confronting this two-dimensional crock of gold? It is currently being developed for laboratory techniques, such as its use as a sieve for separating molecules of different kinds (the production of biofuels is a target) and in desalination (the rendering of seawater potable). Although graphene itself does not readily adsorb gas molecules, its surface—it is almost entirely surface—can be chemically modified to be responsive to gases of different kinds, and their attachment modifies the electrical properties of the underlying graphene sheets so that their presence can be detected.

Chemists naturally wonder whether this two-dimensional wonderland can be inhabited by other materials, and whether those materials can circumvent some of the deficiencies of otherwise seemingly miraculous graphene. New materials of a graphene-like form have been made electrochemically, with compounds like molybdenum sulfide, tungsten sulfide, and more exotic materials based on titanium carbide. Some of these two-dimensional materials show semiconducting properties, which graphene lacks, and have already been fabricated into minute integrated circuits. Graphene itself is open to chemical modification, one procedure being to oxidize it to form graphene oxide. Flakes of this material aggregate into sheets of 'graphene paper' which, materials scientists are hoping, can form the basis of a whole new class of materials with tunable electrical, thermal, optical, and mechanical properties.

New applications

So vast are the applications of the new materials developed by chemists in collaboration with materials scientists, physicists, biologists, and engineers that I can do no more than stand in this Aladdin's cave of wonders and point around at random, knowing that I will miss a crucial development or example, but hoping to convey through just a few examples the impression that life is being transformed by this collaboration.

Thus, I point to self-cleaning glass. This labour-saving development is based on photochemistry and an understanding of the forces of attraction or repulsion between molecules, in particular the property that renders a surface 'hydrophobic', or water repelling. A typical self-cleaning glass is coated with a thin transparent layer of titanium dioxide, which responds to sunlight by breaking down chemically any dirt that happens to be deposited on it. The water-repelling surface means that any water, rainwater in particular, washes away the products of this photocatalysed decomposition without leaving dirty streaks.

I can point to smart fabrics. Smart fabrics can glow with different colours, perhaps representing the wearer's distribution of temperatures and, in a crude way, their emotional state. Or they can respond to the ambient conditions or the whim of the wearer by changing their appearance electrically. Not only must the fabrics be entertaining, they must also withstand the rigours of passage through the laundry and the stress of being worn, crumpled, and creased.

Catalysis is hugely important, as I have already indicated for industry, but is vital for the elimination of pollution from internal combustion engines. The catalytic converter now built into all our cars makes use of some highly sophisticated chemistry, for it must come into operation quickly as soon as the engine is started when it is cold (a significant proportion of pollution occurs then) yet continue to act when the engine is untouchably hot. Moreover, not only must the catalysts achieve reduction of nitrogen oxides to harmless nitrogen, they must also achieve the oxidation of carbon monoxide to carbon dioxide and complete the oxidation of unburned hydrocarbon fuel. Not only that, they also need to respond to the different conditions as the engine runs, such as the leanness or richness of the fuel/air mixture and sudden surges during acceleration. All this needs to be developed by chemists.

Perhaps nowhere is modern chemistry more important than in the development of new drugs to fight disease, ameliorate pain, and enhance the experience of life. *Genomics*, the identification of genes and their complex interplay in governing the production of proteins, is central to current and future advances in pharmacogenomics, the study of how genetic information modifies an individual's response to drugs and offering the prospect of personalized medicine, where a cocktail of drugs is tailored to an individual's genetic composition.

Even more elaborate than genomics is *proteomics*, the study of an organism's entire complement of proteins, the entities that lie at the workface of life and where most drugs act. Here computational chemistry is in essential alliance with medical chemistry, for if a protein implicated in a disease can be identified, and it is desired to terminate its action, then computer modelling of possible molecules that can invade and block its active site is the first step in rational drug discovery. This too is another route to the efficiencies and effectiveness of personalized medicine.

New discoveries

I do not want to give the impression that advances in chemistry are entirely confined to its applications. They are certainly headline-grabbing and affect us all. However, chemists are also engaged in the fundamental business of discovering more about matter and how it may be modified. Increasingly, they are becoming familiar with the workings of Nature at a molecular level, learning her ways, and stumbling on to features that might be astonishing and not have any immediate application except for that most precious of entities, knowledge. Fundamental research is absolutely vital to this endeavour, for it leads on to unforeseen discoveries, unforeseen understanding, and unforeseen applications of extraordinary brilliance.

In order to introduce a certain closing flourish, here I mention a single, singular, particular, purely academic recent discovery: Nature, chemists have discovered, can tie herself into knots. A class of molecules, it has been discovered to the researchers' astonishment and delight, can tie itself spontaneously into a trefoil knot. As a commentator (Fraser Stoddart) on this work remarked 'the new research illustrates some of the finest aspects of synthetic and physical organic chemistry and is one of these rare instances where stereochemistry is being expressed at its most elegant'.

Such is the joy, the intellectual pleasure, that modern chemistry inspires. I hope these pages have erased to some extent those memories that might have contaminated your vision of this extraordinary subject and that you have shared a little of that pleasure.

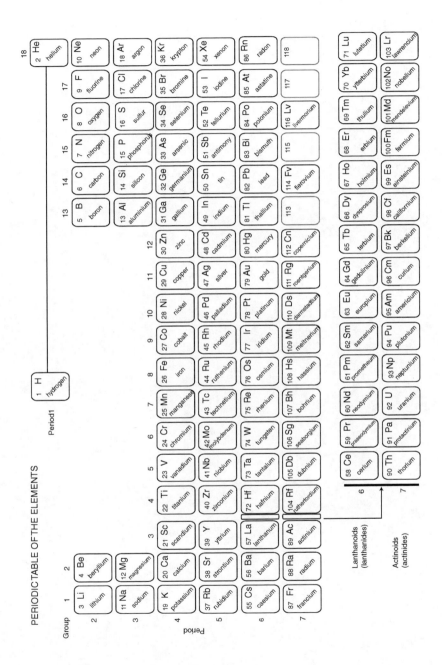

Glossary

Acid A proton donor (see *Lewis acid*).
Alkali A water-soluble base; a solution of a base in water.
Amino acid An organic compound of formula $NH_2CHRCOOH$ (R denotes a group of atoms, such as $-CH_3$, or something more complex).
Analysis The identification of substances and the determination of their amounts and concentrations.
Anion A negatively charged atom or group of atoms.
Atom The smallest particle of an element; an entity composed of a nucleus surrounded by electrons.
Base A proton acceptor (see *Lewis base*).
Bond A shared pair of electrons lying between two atoms.
Carbohydrate An organic compound of typical formula $(CH_2O)_n$.
Catalysis The acceleration of a chemical reaction by a species that undergoes no net change.
Cation A positively charged atom or group of atoms.
Chain reaction A reaction in which a molecule, ion, or radical attacks another, the product attacks another, and so on.
Complex A group of atoms consisting of a central metal atom to which are attached ligands.
Compound A specific combination of elements bonded together.
Diffraction Interference between waves caused by an object in their path.
Double bond Two shared pairs of electrons lying between two atoms.
Electrochemistry The use of chemical reactions to generate electricity and the use of electricity to bring about chemical change.

Electrolysis To achieve a chemical reaction by passing an electric current.

Electron A negatively charged subatomic particle.

Electrophile A species that is attracted to electron-dense (negative) regions.

Electrophilic substitution A substitution reaction in which one reactant is an electrophile.

Element A substance that cannot be broken down chemically into simpler substances; a substance composed of a single kind of atom. For a list of elements and their symbols, see the preceding Periodic Table.

Genomics The identification of genes and their complex interplay in governing the production of proteins.

Green chemistry The aim to minimize the impact of chemical manufacturing processes on the environment by strict guidelines about the use of materials and the elimination of waste.

Hydronium ion H_3O^+.

Hydroxide ion OH^-.

Intermediate See *Reaction intermediate*.

Ion An electrically charged atom or group of atoms (see *Cation* and *Anion*).

Isotopes Atoms with nuclei with the same atomic number (number of protons) but different numbers of neutrons.

Lewis acid An electron pair acceptor.

Lewis base An electron pair donor.

Lewis acid–base reaction A reaction of the form $A + :B \rightarrow A-B$ between a Lewis acid and a Lewis base.

Ligand A group of atoms attached to a central metal atom in a complex.

Lone pair A pair of electrons not involved directly in bond formation.

Mixture A mingling of substances without the formation of new chemical bonds.

Molecule The smallest particle of a compound; a discrete combination of atoms in a definite arrangement.

Monomer A small molecule used in a polymerization reaction.

Nucleophile A species that is attracted to electron-poor (positive) regions.

Nucleophilic substitution A substitution reaction in which one reactant is a nucleophile.

Oxidation The removal of electrons from a species; reaction with oxygen.

Photon A particle of electromagnetic radiation.

Polymer The product of a polymerization reaction.
Polymerization The linking together of small molecules to create long chains.
Product The material produced by a chemical reaction.
Protein A complex compound built from amino acids.
Proteomics The study of an organism's entire complement of proteins.
Proton The nucleus of a hydrogen atom.
Radical A species with at least one unpaired electron.
Reactant The starting material in a specified chemical reaction.
Reaction intermediate A species other than the reactants and products that is proposed to be involved in a reaction mechanism.
Reagent A substance used as a reactant in a variety of chemical reactions.
Redox reaction A reaction involving oxidation of one species and reduction of another; an electron transfer reaction.
Reduction The addition of electrons to a species.
Salt An ionic compound formed by the reaction of an acid and a base.
Solute A dissolved substance.
Species Used here to denote an atom, molecule, or ion.
Spectroscopy The observation of the absorption or emission of radiation by a sample.
Substitution reaction A reaction in which an atom or group of atoms is substituted for one already present in a molecule.
Superconductivity The ability to conduct electricity without resistance.
Synthesis The creation of substances from simpler components.
Titration The determination of the concentration of an acid (or base) by measuring the volume of an alkali (or acid) needed to neutralize it.
Transition metal A member of Groups 3 to 11 of the Periodic Table.

Further reading

For a survey of chemical thermodynamics, see my *Four Laws that Drive the Universe* (2007), reissued as *The Laws of Thermodynamics: A Very Short Introduction*, Oxford: Oxford University Press (2010).

For a broader introduction to the principles of chemistry, see my *Physical Chemistry: A Very Short Introduction*, Oxford: Oxford University Press (2014).

The variety of chemical reactions that is merely touched on in this volume is elaborated pictorially in my *Reactions: The Private Life of Atoms*, Oxford: Oxford University Press (2011).

For a broad survey of the principles and techniques of chemistry see my *Chemical Principles: The Quest for Insight*, with Loretta Jones and Leroy Laverman, New York: W.H. Freeman & Co (2013).

Others, of course, have written wonderfully and extensively on chemistry. The *Very Short Introductions* are particularly apposite, and include

Molecules, Philip Ball (2003)

The Elements, Philip Ball (2004)

The Periodic Table, Eric Scerri (2011)

For a survey of modern trends and applications of chemistry, see *The New Chemistry*, ed. Nina Hall, Cambridge: Cambridge University Press (2000).

"牛津通识读本"已出书目

古典哲学的趣味　　福柯　　　　　　　地球
人生的意义　　　　缤纷的语言学　　　记忆
文学理论入门　　　达达和超现实主义　法律
大众经济学　　　　佛学概论　　　　　中国文学
历史之源　　　　　维特根斯坦与哲学　托克维尔
设计，无处不在　　科学哲学　　　　　休谟
生活中的心理学　　印度哲学祛魅　　　分子
政治的历史与边界　克尔凯郭尔　　　　法国大革命
哲学的思与惑　　　科学革命　　　　　民族主义
资本主义　　　　　广告　　　　　　　科幻作品
美国总统制　　　　数学　　　　　　　罗素
海德格尔　　　　　叔本华　　　　　　美国政党与选举
我们时代的伦理学　笛卡尔　　　　　　美国最高法院
卡夫卡是谁　　　　基督教神学　　　　纪录片
考古学的过去与未来　犹太人与犹太教　大萧条与罗斯福新政
天文学简史　　　　现代日本　　　　　领导力
社会学的意识　　　罗兰·巴特　　　　无神论
康德　　　　　　　马基雅维里　　　　罗马共和国
尼采　　　　　　　全球经济史　　　　美国国会
亚里士多德的世界　进化　　　　　　　民主
西方艺术新论　　　性存在　　　　　　英格兰文学
全球化面面观　　　量子理论　　　　　现代主义
简明逻辑学　　　　牛顿新传　　　　　网络
法哲学：价值与事实　国际移民　　　　自闭症
政治哲学与幸福根基　哈贝马斯　　　　德里达
选择理论　　　　　医学伦理　　　　　浪漫主义
后殖民主义与世界格局　黑格尔　　　　批判理论

德国文学	儿童心理学	电影
戏剧	时装	俄罗斯文学
腐败	现代拉丁美洲文学	古典文学
医事法	卢梭	大数据
癌症	隐私	洛克
植物	电影音乐	幸福
法语文学	抑郁症	免疫系统
微观经济学	传染病	银行学
湖泊	希腊化时代	景观设计学
拜占庭	知识	神圣罗马帝国
司法心理学	环境伦理学	大流行病
发展	美国革命	亚历山大大帝
农业	元素周期表	气候
特洛伊战争	人口学	第二次世界大战
巴比伦尼亚	社会心理学	中世纪
河流	动物	工业革命
战争与技术	项目管理	传记
品牌学	美学	公共管理
数学简史	管理学	社会语言学
物理学	卫星	物质
行为经济学	国际法	学习
计算机科学	计算机	化学